Evaluation of Science
and Technology Education
at the Dawn
of a New Millennium

T0235820

INNOVATIONS IN SCIENCE EDUCATION AND TECHNOLOGY

Series Editor:

Karen C. Cohen, Harvard University, Cambridge, Massachusetts

A Continuation Order Plan is available for this series. A continuation order will bring delivery of each new volume immediately upon publication. Volumes are billed only upon actual shipment. For further information please contact the publisher.

Evaulation of Science and Technology Education at the Dawn of a New Millennium

Edited by

James W. Altschuld

The Ohio State University
Columbus, Ohio

and

David D. Kumar

Florida Atlantic University
Davie, Florida

Kluwer Academic/Plenum Publishers
New York, Boston, Dordrecht, London, Moscow

Library of Congress Cataloging-in-Publication Data

Evaluation of science and technology education at the dawn of a new millennium/edited by James. W. Altschuld and David D. Kumar.
    p. cm.
    Includes bibliographical references and index.
    ISBN 0-306-46749-6
      1. Science—Study and teaching—United States—Evaluation. 2. Technical education—United States—Evaluation I. Altschuld, James W. II. Kumar, David D.

Q183.3.A1 .E79 2002
507'.1'0973—dc21

2001057962

ISBN: 0-306-46749-6

©2002 Kluwer Academic/Plenum Publishers
233 Spring Street, New York, NY 10013

http://www.wkap.nl/

10   9   8   7   6   5   4   3   2   1

Printed in the United States of America

Foreword

James Altschuld, David Kumar, and their chapter authors have produced an upbeat, provocative, visionary, and useful volume on educational evaluation. Of special utility is its grounding in issues and practices relating to evaluations of science and technology education. The book should appeal and be useful to a wide range of persons involved in evaluations of educational policy, programs, and (less so) science teachers. These persons include science and technology education experts, educational policymakers, officials of the National Science Foundation, school administrators, classroom teachers, evaluation instructors, evaluation methodologists, practicing evaluators, and test developers, among others. Contents reflecting international studies of curriculum, evaluation of distance education, and evaluation of technology utilization in Australian schools as well as evaluations in America should make the book appealing to an international audience. Moreover, it provides a global perspective for assessing and strengthening educational evaluation in the U.S.

The book argues convincingly that sound evaluation is needed to improve science and technology education and strengthen its claim on public resources. The book speaks thoughtfully and appropriately circumspectly to the efforts promulgated by the National Science Foundation to employ process and outcome drivers to evaluate systemic and other more specific reforms in science education. Lessons from efforts to evaluate uses of technology in schools provide insights into the complexities of technology-based instruction and the challenges and promise of using evaluation to make technology a more effective, compatible part of science teaching. The book addresses, albeit minimally, science teacher evaluation as well as program and policy evaluation. The book examines the well-known TIMSS as an example of how international evaluations of curriculum can affect thinking about and also influence national education policy. TIMSS is also suggested as a basis for developing better, curriculum-specific tests.

In its discourse covering a wide range of recently developed evaluation techniques, the book shows that educational evaluation has been making steady progress toward achieving the status of an area of effective professional service. Throughout the volume are repeated emphases on the importance of employing multiple evaluation methods, employing qualitative as well as quantitative techniques, and triangulating findings to more confidently assess educational programs. (This message is reminiscent of Ralph Tyler's almost unknown, but wonderful writing in the same vein about 70 years ago in his 1932 Ohio State University-published book, *Service Studied in Higher Education.*) The book's authors agree that employment of multiple methods is still a difficult, costly, matter and that it often requires team efforts. Another implication of this book clearly is that evaluation training programs should provide instruction and practical experiences in a wide range of evaluation procedures for teachers as well as evaluation specialists.

While mindful and respectful of progress made in developing the still fledgling evaluation field, this book is no celebration of evaluation's status quo. Examples of effective evaluation in science and technology are identified as scarce. Several chapters grapple with the ubiquitous, troubling area of student testing, reminding that high stakes testing is often counterproductive and that there has been too little progress to make student testing a positive force for improving teaching and learning. The political orientation of large-scale testing is also discussed to help explain why highly questionable accountability-oriented testing programs have survived, despite their validity deficiencies and many counterproductive impacts on schools, teachers, and students. A key, sensible position in the chapter on Assessment Reform is that student testing is most effective when integrated in curriculum and classroom teaching and that much more effort should be devoted to this promising, but underdeveloped application of achievement testing, including better and more use of performance tests. The book also illustrates the problems of language in the educational evaluation field, with many terms having unclear or different meanings in the different chapters.

Collectively, the authors look to the future of what might be in the educational evaluation realm. They challenge evaluation theorists, methodologists, and practitioners to be more creative and productive. They counsel these groups to cast aside the failed or marginally useful evaluation procedures of the present and past and systematically to pursue better ways to conduct and use evaluation to improve teaching and learning. The book appropriately stresses that evaluators should devote much more attention to context and process and notes that each is highly complex, yet crucially influential on the effectiveness of programs and the success of individual teachers and other educators. A critically important point is made, under the label of context, that evaluators must be sensitive to the needs, characteristics, and dynamics of different levels of schooling. Information needs do vary considerably across classroom, school, school district, state, and national levels; evaluation plans, processes, and reports should take these differences into account if evaluations are to be useful at each level. To the point of improving teaching and learning, there is discussion of the pros and cons of classroom- and school-based, improvement-oriented evaluations, as contrasted to the more familiar external, summatively-oriented, and presumably more objective evaluations.

This book is a provocative read, and it contains much practical advice. While the different chapters hold together well, they present no party line. Readers will find contrasting views and orientations across the chapters as well as common themes. The writing reflects the realities and complexities of evaluation in education. The different, sometimes conflicting messages point up needs for better, clearer conceptualizations of evaluation and pertinent research.

Practitioners and theorists alike should come to prize this book as a valuable, user-friendly resource. The initial chapter provides a functional overview, analysis, and guide to the ensuing chapters. Many readers likely will repeatedly revisit certain chapters to examine particular evaluation issues. Practitioners will probably appreciate the inclusion of practical evaluation forms. Evaluation methodologists should find the book a useful repository of evaluation issues and promising practices as they contemplate how best to improve evaluation approaches and methods. Leaders at the National Science Foundation, state education departments, education service organizations, charitable foundations, and school districts should find the book a good resource as they consider how to increase and improve evaluations in science, mathematics, and technology education. Teachers and department heads no doubt will find much practical advice they can employ in classrooms, especially concerning the appropriate use of testing to assess and strengthen student learning. The book should also be useful in evaluation courses at both undergraduate and graduate levels. While the book helpfully focuses on evaluations in science, mathematics, and technology education, many of its lessons and the dilemmas it surfaces are applicable to a wide variety of education contexts. I congratulate the editors and authors on making a valuable contribution to the literature of educational evaluation.

Daniel L. Stufflebeam

Preface

Evaluation is in a state of flux. Perspectives of what it is, what it has been, and where it is going are changing, in general, and in science and technology education, in specific. Evaluation is no longer just a service or process, it is in the realm of a 'near discipline' (Worthen, 1994) or what has been referred to as 'trans-discipline' (Scriven, 1994). Evaluation has matured as a field as indicated by the existence of professional evaluation groups (the American Evaluation Association, and international, and state associations) and many high quality journals that deal with evaluation issues and research.

In terms of methodology, evaluation has evolved in dramatic fashion. Evaluators now include in their tool boxes concept mapping, evaluability assessment, tree diagrams, alternative and authentic assessment, logic models, and numerous methods for assessing needs. They have also moved far beyond the methodological focus to the consideration of other concepts such as the impact of contextual factors on programs and how findings are utilized to make decisions.

Another example of change in evaluation is the increased demand for accountability that is evident in statewide testing programs, the release of school achievement results to the general public, and the Government Performance and Results Act (GPRA). Collectively, there is growing pressure on educators to carefully evaluate both the outputs and outcomes of their programs. Science education is under the same intense public scrutiny as other areas of education (e.g., technology) and will increasingly need evaluation to help make its case in the public arena.

The background just described prompted our decision to undertake this book. We saw the beginning of the third millennium as an opportunity to explore the current state of evaluation in science education and technology, and what lies ahead for it.

GOAL OF THE BOOK AND INTENDED AUDIENCES

The goal of this book is to examine how evaluation has functioned in the past and the potential and exciting roles it could play in the future for science education and technology. The book deals with the following overarching themes in evaluation; Evaluation of Reform Efforts, Evaluation of Science, Evaluation of Technology; Evaluation of Process and Achievement; Evaluation of Teacher Training Programs; Policy; Research and Methodology.

The book is intended for a number of different yet overlapping audiences. First and foremost would be evaluators charged with providing information about the implementation and results of science and technology education programs and individuals who teach evaluation at universities. We believe that our choice of authors and topics produced extremely challenging ideas and very useful information for these two groups. The second audience consists of policy analysts, decision-makers, and legislators who are actively engaged in making educational choices for the future. How programs have been evaluated, how they will be evaluated in the future, and the nature of evaluation results should be of great value to this type of audience. The third constituency consists of teachers, curriculum specialists, and department heads in schools and school systems who have to make some of the hard choices regarding science and technology education.

OUR CHARGE TO THE CHAPTER AUTHORS

Authors were chosen based upon their experience, knowledge, and publications in science education and evaluation. A large number of them are immediately recognizable as leaders in science and technology education and/or evaluation. Others will become so in the near term.

In devising chapter outlines, we could have been prescriptive and controlling. That would have been the worst thing to do and would have defeated our intent. Instead, we tried to capitalize on the strengths of our authors by sparking their imaginations and encouraging them even to the point of 'going out on a limb.' We think we succeeded but, of course, not us but you, the readers of this book, are the ultimate, unbiased judges as to whether or not such perceptions are warranted.

The general instructions given were to consider what had been done in evaluation in the specific focus of the chapter and then to speculate on what the future might hold for evaluation. In some cases, the authors spring-boarded from our admittedly brief guidelines into wholly new, unanticipated, and exciting directions. We are pleased that they did so and that they had the confidence in us to exercise this freedom. We sincerely hope that you agree that they have pushed the creative envelope.

ORGANIZATION OF THE TEXT

The book begins with a Foreword written by Daniel Stufflebeam of the Evaluation Center at Western Michigan University. Stufflebeam has been an internationally recognized leader in the field of evaluation for more than 35 years. Aside from his faculty duties he has just added an entire issue of New Directions in Evaluation (2001) to his extensive list of evaluation publications. It is devoted to a thoughtful examination of evaluation models. He currently serves on the Board of the American Evaluation Association.

CHAPTER 1
WHAT DOES THE FUTURE HAVE IN STORE FOR THE EVALUATION OF SCIENCE AND TECHNOLOGY EDUCATION?

James Altschuld and David Kumar (of The Ohio State University and Florida Atlantic University, respectively) analyze and summarize the ideas and concepts produced by the knowledgeable and well versed slate of chapter authors. Rather than placing a chapter like this at the end of the book as is usually done, it is better up front. It is derived from an analysis of all of the chapters especially in regard to important themes that seemed to cut across them. For example: what can we learn from the chapter authors? what are the challenges facing science and technology education and how will evaluators collect and provide information that is useful for program improvement? what guidance can we offer a new generation of evaluators?

CHAPTER 2
WHAT ROLE SHOULD TIMSS PLAY IN THE EVALUATION OF U.S. SCIENCE EDUCATION?

William Schmidt and HsingChi Wang (of Michigan State University) explore the role of TIMSS (Third International Mathematics and Science Study) in evaluation. In particular, they view TIMSS as a mechanism for thinking about the nature of the U.S. system of education as a whole and for influencing policymaking bodies.

CHAPTER 3
EVALUATING SYSTEMIC REFORM: EVALUATION NEEDS, PRACTICES, AND CHALLENGES

Bernice Anderson of the National Science Foundation has informed us of the complexity and subtlety of evaluating major science education reform endeavors. Her chapter deals with a sobering reminder of just how difficult it is to define what we mean by reform, to develop strategies and models to guide evaluations, to specify the critical variables to be studied, and to carry out the overall evaluations. The chapter is rich in examples and illustrates the fact that evaluating the success or failure of reform is filled with exciting challenges yet daunting at the same time.

CHAPTER 4
MUSINGS ON SCIENCE PROGRAM EVALUATION IN AN ERA OF EDUCATIONAL ACCOUNTABILITY

Dennis Cheek of the Rhode Island Department of Education and the University of Rhode Island discusses a view of accountability systems as technologies of social control. He then proceeds to describe both the assumptions and reasons for testing as well as the design flaws in state testing programs and the impact of testing programs at the school level. Lastly, he argues for better assessment of student achievement.

CHAPTER 5
ASSESSMENT REFORM: A DECADE OF CHANGE?

Wendy McColskey of SERVE (Southeast Regional Vision for Education located in Greensboro) and Rita O'Sullivan of the University of North Carolina (Chapel Hill) are evaluators who have been training teachers (especially science teachers) in developing and using alternative assessment procedures in their classrooms. Their chapter looks at the national and state level tests, but then rapidly moves to the classroom as the main arena for action especially in regard to teacher training for assessment. It is in the classroom that meaningful, more authentic assessment takes place with the ensuing result that teachers become more reflective learners about their own teaching efforts. This counterpoint should make all of us more reflective of evaluation in science education.

CHAPTER 6
EVALUATION OF INFORMATION TECHNOLOGY

John Owen, Gerard Calnin, and Faye Lambert of the University of Melbourne in Australia have helped us to understand the difficulty of evaluating what seems like a straightforward and simple undertaking, that of introducing technology (information technology) into classrooms. They show us that there are problems in defining the construct (it has many meanings) as well as evaluating its implementation and effectiveness. They also identify the need to delve into the contexts of the classroom and school to operationalize an information technology instructional system in science education.

CHAPTER 7
COMPLEMENTARY APPROACHES TO EVALUATING TECHNOLOGY IN SCIENCE EDUCATION

David Kumar of Florida Atlantic University and James Altschuld of The Ohio State University primarily look at two ways of evaluating a teacher training program that relied on the extensive use of technology. One group of evaluators (Barron et al., 1993) used what might be termed standard methods for conducting an evaluation, whereas, the other applied a context evaluation framework to the same situation. The two evaluations generated quite distinct yet complementary results and helped to raise a number of questions and issues about how technology intensive projects could be evaluated.

CHAPTER 8
EVALUATION OF SCIENCE TEACHING PERFORMANCE THROUGH COTEACHING AND COGENERATIVE DIALOGUING

Kenneth Tobin and Wolff-Michael Roth (of the Universities of Pennsylvania and British Columbia, respectively) ask us to reexamine the basis of traditional thought regarding how science teachers are evaluated and trained. They offer a strikingly different view as to how to think about the evaluation of novice teachers and the procedures normally employed in such evaluations. Their intensive approach to both training and evaluation are totally unique when compared to what is presently done in almost all teacher training institutions. As participants in the process, they use the novelty of dialogue to provide a rich description of intimate, up-close feelings and tribulations about what takes place in the classroom.

CHAPTER 9
EVALUATING SCIENCE INQUIRY: A MIXED METHOD APPROACH

Douglas Huffman of the University of Minnesota examines evaluation of science education from an inquiry perspective based on his experience in research and study in this area. The topic is particularly important as much of the science education literature stresses process, but with little explanation/definition of what is meant by the word or how the elusive beast can be or is evaluated. Huffman presents a model to guide evaluation in this regard and a concrete example from his work.

CHAPTER 10
DISTANCE LEARNING IN SCIENCE EDUCATION:
PRACTICES AND EVALUATION

John Cannon of the University of Nevada at Reno illuminates for us that distance education programs in general and science education in specific have not been often or systematically evaluated. He laments this current status in light of the perception that distance education will greatly affect science education delivery and instruction in the near future as suggested by the fact that more than 17,000 total distance education courses were esti-mated to be available in the year 2000. Cannon described many of the issues beneath the surface of evaluation in distance education even noting that we apply, perhaps inappropriately, an evaluation methodology designed for traditional courses to a new medium.

SUGGESTION TO THE READER

We have enjoyed our journey in preparing this book and in working with the chapter authors. We invite you to share that exhilaration as you read their ideas and expectations for the future. Have as much fun as we did! If that excitement captures you please feel free to contact us or the chapter writers with your comments and thoughts.

Contents

CHAPTER 1

What Does the Future Have in Store for the Evaluation of Science and Technology Education?

James W. Altschuld and David D. Kumar

> A hundred years ago, as the nineteenth century drew to a close, scientists around the world were satisfied that they had arrived at an accurate picture of the physical world. As physicist Alastair Rae put it "By the end of the nineteenth century it seemed that the basic fundamental principles governing the behavior of the physical universe were known" (Rae, 1994). Indeed, many scientists said that the study of physics was nearly completed: no big discoveries were to be made only details and finishing touches.
>
> Michael Crichton from the introduction to the 1999 science fiction book, *Timeline*

This introduction to *Timeline* draws a clear parallel to work in the evaluation of science and technology education programs. To begin this

James W. Altschuld, Educational Policy and Leadership, The Ohio State University, 29 W. Woodruff Avenue, Columbus, OH 43210.   David D. Kumar, College of Education, Florida Atlantic University, 2912 College Avenue, Davie, FL 33314.

Evaluation of Science and Technology Education at the Dawn of a New Millennium, edited by James W. Altschuld and David D. Kumar, Kluwer Academic / Plenum Publishers, New York, 2002.

1

journey, we drafted a prospectus that looked as though it could become a catalogue of basic and well known parameters of evaluation—perhaps a bit pedestrian, but nevertheless a useful compendium of important knowledge. As the 19th Century scientists were hopelessly wrong so were we.

The prospectus for the book was very general in nature permitting flexibility in approaches to assigned topics. To our good fortune and that of you, the readers, chapter authors would not let us get away with such modest initial thinking, they just ran away with ideas. The authors collectively demonstrate that the need for evaluation in science and technology education continues to be enormous. They consistently identify aspects of these two areas in which saying that there is limited evaluation taking place would be a serious overstatement. They describe a literature-base that is frequently and even frighteningly sparse with great opportunities for major breakthroughs in how to deliver programs as well as how to evaluate them. They concretely illustrate the necessity of carefully defining and describing the entity to be evaluated while at the same time making us aware of the difficulty in regard to the "under the surface" (almost hidden) aspects of science process, information technology, and systemic reform to name a few topics.

Science and technology education are in reality complex phenomena embedded in multifaceted, intense, and complicated educational contexts. Identifying, separating out, and isolating features of programs and their effects is not easily done. Methodologically, the authors are virtually emphasizing the use of multiple sources of data and multiple methods to provide documentation regarding how innovative programs are being implemented and in determining their outcomes. Moreover, they routinely employ, either explicitly or implicitly, theoretical models or frameworks that guided their efforts thereby underscoring the importance and the influence of such structures on the design and conduct of evaluations. The chapters are exciting and open new possibilities for carrying out evaluations, and for conducting research on evaluation in addition to the substantive dimensions of science and technology education.

In edited volumes such as this one, a summarizing chapter written by the editors is usually the last one in the text. In this case the order is reversed, an action in accord with the original plans for the book. What propelled our thinking, why did we choose to go this way? The reasoning was as follows. All of the chapter authors, except for the two co-editors, did not know what the others were writing and in most cases they were not personally acquainted with each other. They were given a general topic and were free to deal with it in any manner they chose. Suppose that from this nearly orthogonal work, similar issues and themes were generated without understanding or prior knowledge of what others were doing. Would their independent yet shared ideas not be a form of verification of trends and important emphases for consideration?

If this were to occur, and it did, then what emerges is a highly interesting data set consisting of the perceptions of very experienced and knowledgeable individuals that could be analyzed in an informal qualitative manner. Are there similar issues and variables embedded in the chapters? What concerns are being expressed for the evaluation of science and technology education? What are the potential needs for the field suggested by the authors? What subtle concerns are affecting these types of educational programs and their evaluations. Metaphorically the authors are speaking to and telling us, and, in turn you, a very intriguing story.

With that logic in mind, the methodology that we applied is similar to an across-evaluation-model comparison found in Worthen, Sanders, and Fitzpatrick (1997) and an across chapter analysis carried out by Kumar and Chubin (2000) in regard to their Science-Technology-Society book (2000). It is a version of constant-comparative analysis like that described in Chapter 7. It is a powerful tool and, in this instance, one that leads to highly utilitarian results. At the same time we make no pretense of having captured anything more than a few of the overarching constructs contained in the chapters. We only present highlights from the many ideas proposed by the authors. The responsibility for going beyond our spanning themes ultimately resides with the readers of the text.

Here is what we did. First, the points made in each chapter were studied to determine the main features, constructs, and variables that were being stressed. Next we proceeded to a more detailed examination of some of the deeper thoughts and ideas embedded in the chapters. Then, explanatory themes were postulated that cut across the chapters and provided a cohesive link for the common elements found within them. Lastly, suggestions for the future of science and technology education evaluation emanating from the prior steps are provided. Although the chapter authors were charged with dealing with a single focused topic they went far beyond that simplistic beginning and, again, we sincerely hope that the explanatory themes inferred from their efforts are accurately drawn and honestly and fully portray the richness of their thoughts.

OVERVIEW OF THE CHAPTERS

In Table 1, an overview of the chapters is given. It is obvious even from the brief commentary in the table that the chapters have an immense amount of information and insights on a wide range of content about the evaluation of science and technology education.

Table 1. Analysis of the Main Concepts of the Chapters

Chapter No., Title, Author(s)	Main Idea	Special Features
1. What Does the Future Have in Store for the Evaluation of Science and Technology Education? James W. Altschuld & David D. Kumar	Examination of the other chapters with a view toward emerging themes, ideas that cut across them, and guidance in thinking about the future of evaluation in science and technology education.	Themes emerged that were repeatedly observed in chapters or that were easily inferred during the review process.
2. What Role Should TIMSS Play in the Evaluation of U. S. Science Education? William Schmidt & HsingChi Wang	How should TIMSS affect views about science education in the United States? The conclusion drawn is that TIMSS relates to the formation of national not local policy, especially since there is greater variation across countries than across school districts in the U.S.	The drop in performance of U.S. students from the 4th to the 8th grades is noted. The authors look at how the curriculum is organized for the teaching of science as contrasted to other countries.
3. Evaluating Systemic Reform: Evaluation Needs, Practices, and Challenges Bernice Anderson	A comprehensive picture is given of how various federally funded (primarily by The National Science Foundation), systemic reform efforts have been or are being evaluated in the United States.	The chapter is particularly insightful in terms of models or structures to focus the evaluation of complex reform endeavors and the subtle issues that need to be considered in such evaluations.
4. Musings on Science Program Evaluation in an Era of Educational Accountability Dennis Cheek	After raising a number of concerns about the underlying validity of accountability programs, it is posited that such programs are mechanisms for asserting power and control of the state over local school districts.	Design flaws of testing programs are described, better ways of assessing students are suggested, and the observation is made that meaningful change in science teaching and learning will only occur at the classroom level.
5. Assessment Reform: A Decade of Change Wendy McColskey & Rita O'Sullivan	Are accountability programs promoting student learning and achievement? Data are presented that suggest that accountability programs are not accomplishing such goals. Change leading to improved results, while facilitated by school systems, must reside at the classroom and teacher levels.	Discussion centers on promoting higher order learning via alternative and authentic assessment techniques; training for teachers who have had little formal exposure to assessment methods; and greater involvement of the local level in accountability conversations.
6. Evaluation of Information	Information technology is a generic	Stress is placed on the idea that using

Technology
John Owen, Gerard Calnin, & Faye Lambert

concept that can take many shapes and forms when used in classrooms. If fully integrated into science education as well as classrooms in other subjects, information technology may be a powerful tool for enhancing learning. information technology in a deeply integrated, well articulated manner is by far the exception rather than the rule.

7. Complementary Approaches to Evaluating Technology in Science Education
David Kumar & James Altschuld

Two unique ways of evaluating an interactive media teacher education project are depicted as well as the strikingly different results that were produced via their implementation. The lack of literature citations for the evaluation of technology in science education, particularly in relationship to pre-service programs, is noted. The fact that the two evaluations used different methodologies is underscored as indicative of the need for mixed methods in evaluating projects. The demands of employing mixed methods are emphasized.

8. Evaluation of Science Teaching Performance Through Co-teaching and Co-generative Dialoguing
Kenneth Tobin & Wolff-Michael Roth

Evaluation is typically judgmental when it comes to looking at the performance of teachers with the evaluation of science teachers conforming to this general framework. A virtual dichotomy is established with the evaluator being apart from the situation, which the authors suggest does not promote meaningful evaluation. The evaluator cannot be apart from the person and situation being evaluated and, to a high degree, must become a co-teacher or co-participant. Being an outside judge is not advocated and strongly de-emphasized in the chapter.

9. Evaluating Science Inquiry: A Mixed Method Approach
Douglas Huffman

Evaluating the process of science inquiry requires that the construct be defined. Various definitions are reviewed followed by the construction of a three-level model the inquiry process used to focus an of evaluation conducted by the author and others. The model and its utilization are explained in depth. The chapter contains numerous samples of forms that were employed in the evaluation.

10. Distance Learning in Science Education: Practices and Evaluation
John Cannon

After demonstrating the rapid growth in web based learning opportunities, the paucity of evaluation studies examining how they have been implemented and the results they have produced is lamented. Multiple delivery forms of web based instruction are reviewed. A key feature here is the concept that even when distance education is evaluated, these evaluations tend to be traditional in nature instead of being adopted to the dimensions of the medium used for the delivery of instruction.

A second, more subtle outcome from the table relates to common concepts. The overall purpose of the book is to look at the evaluation of science and technology education, hence all the chapters would be expected to deal with this emphasis in a prominent manner and certainly they did so. As pointed out before, they additionally focus on a host of other similar if not identical concepts. A few illustrations will show this point.

Chapters 3, 7, and 9, in particular, directly stress via discourse or case-study examples, the use of multiple methods for the evaluation of educational programs, despite the fact that there are a myriad of problems (costs, skills of the evaluator with different methods, questions about whether the methods do or don't fit together) associated with such use. A number of the chapters (4 and 5, in particular) are concerned with the efficacy of accountability systems and the degree to which they will lead to not just change in systems, but more importantly meaningful learning and altered and improved classrooms. These ideas generalize from the science classroom to almost any educational context. The underlying premises might be "what do such systems do", "are such systems really doing any good", and perhaps even more importantly "can they do any good?" This type of questioning stance is found in most chapters. As we proceed through the analysis many other examples of common elements will be described and discussed.

A CLOSER LOOK AT CHAPTER CONTENT

To continue the analysis and move forward from Table 1, additional questions were asked of the chapters based upon an in-depth reading of them and the original summarized notes. They were more closely probed in relation to the following kinds of concerns:

— Is there any evidence of the use of evaluation models or other types of models/theoretical approaches and structures to facilitate work in evaluation? And, if so, what did they look like and how prominent were they?;
— What kinds of methods are employed in the evaluations?;
— Are the authors considering the role and influence of context on the delivery of instruction and how was that influence evaluated?;
— How do they handle defining some of the complex constructs that are embedded in the evaluation of science and technology education programs and projects?; and
— What advice do they offer for the future of evaluation?

Thus, Table 2 was developed. In it, when the issue of **models** and their roles in guiding both development and evaluation are studied, two different types are observed. Certainly, an evaluation model would be evident in all the chapters that focused on testing and accountability. Implicitly or explicitly, there is an accountability framework or assumption running through these chapters. Another evaluation model is found in Chapter 7. It delves specifically into characteristics of environments that facilitate change and improvement or by their absence constitute a non-supportive and, to some degree, a debilitating environment that will not enhance new programs and initiatives.

The other type of model deals with the substantive nature of the topic in a chapter. Examples abound. One is that there are process and outcome drivers that must be in place for reform to be meaningfully implemented and eventually take hold within a setting (see Chapter 3). These drivers become the theoretical basis for a program upon which its evaluation design could be predicated. (Additionally, it might be argued that the drivers are really part of the dimensions of context evaluation as posited in Chapter 7.)

The evaluations in Chapters 6 and 9 follow a similar substantive line of attack. Theoretical models are proposed for the evaluation of information technology and inquiry processes, respectively. Chapter 6 includes a framework of graduated levels regarding the use and integration of information technology into classrooms. That framework ranges from technology just being there with lip service being given to its full incorporation into the classroom in terms of well thought out learning and instructional activities.

Whether they are formally stated or a more emergent kind of imbedded, situation specific program/project theory or entity (and possibly not formally addressed), evaluators normally seek to identify or create such structures to guide their efforts. In that sense there is nothing new in the chapters. The notion of theory as the bedrock of evaluation has been explicated by Chen (1990) and is indirectly supported in evaluation through the application of evaluability assessment procedures (Wholey, 1987). Logic models especially as ways to think about and plan the eventual evaluation of programs are also pertinent here.

From a skeptical point of view, as just noted, there is not much new or of value in the observation about theory in this context. On the other hand, Altschuld and Kumar (1995) conducted a 25 year review of the literature in science education evaluation and found minimal mention of evaluation theories, models, or structures, in particular. At least in the key word search they used for that literature review, science educators were not referring to models. It seemed then that talking about or following models or model-like structures was a moribund type of idea. Now, from the

Table 2. Results of the In-Depth Probing of the Chapters

Chapter	Use of Models	Methodology Focus	The Role of Context	Definition Considerations	Ideas about the Future
1. What Does the Future Have in Store for the Evaluation of Science and Technology Education?	Models of two types (evaluation, substantive) are in Chapters 3, 6, 7, and 9. Models are implied in a number of other chapters.	Mixed methodology is either used or implicit in many of the chapters.	Context is the name of the game. Many chapters focus on the macro (large system) context or imply same. Some (#s 7, 8) deal with what might be termed the micro context.	'Hard-to-define' terms and constructs are apparent throughout the text.	See the final part of the chapter for a brief sampling of what the future might hold for the evaluation of science and technology education.
2. What Role Should TIMSS Play in the Evaluation of U.S. Science Education?	Use of a traditional large-scale testing model with its outcome focus.	While testing is emphasized, the analysis of course offerings and curricula is more qualitative in nature.	The context is investigated in terms of how courses and their organization may affect test outcomes.	Many subtle concepts such as policy influence, curriculum coherence and conceptual integration are imbedded in the text.	The chapter concludes with ideas for looking at curriculum specific tests, isolating learning by grade levels, and other current TIMSS related studies.
3. Evaluating Systemic Reform: Evaluation Needs, Practices, and Challenges	Structure of forces/ factors (drivers) that promote reform. There are two categories of drivers—those that affect **process** and those related to **outcomes**.	Different studies of reform are described. No one methodology predominates and the application of mixed methods occurs often and fits well for the evaluation of reform efforts.	The drivers are, in essence, features or parameters of the context. Examples are given of studies related to factors that promote reform.	Maintaining clear distinctions in light of the complexity of reform initiatives is difficult. See for example, the acronyms (SSI, USI, RSI, and USP) and the many evaluation studies conducted or underway.	Future challenges are related to mixed methodology, leadership of reform, equity issues, and longitudinal analyses of data.
4. Musings on Science Program Evaluation in an	The focus is on testing and accountability	The use of mixed methods is posited as a challenge to	By inference, the context of the overall system is	Different terms may be used for similar concepts when	The call is for more effort in regard to testing and evaluation

Era of Educational Accountability	systems and their use in power disparity (a political model).	the evaluation community.	paramount but with recognition that change also has to be at the 'grass roots' level.	compared to other authors, e.g. curriculum alignment vs coherence integration in Chapter 2.	focused at the 'action' research level of teachers, administrators, and parents. Schools need help.
5. Assessment Reform: A Decade of Change	Thinking about reform at the classroom level is used as a contrasting point to an accountability type of approach (model).	By virtue of scoring rubrics for alternative assessment, multiple methods are either explicitly discussed or inferred.	The classroom context is prime in this discussion particularly as related to teachers' knowledge and understanding of assessment.	Alternative assessment is more understandable through examining its implementation. It is hard to use formal definitions or to impose limits on such situations.	Similar to the prior chapter, the emphasis is on the classroom and local levels becoming more involved in the accountability debate and influencing the quality of the curriculum.
6. Evaluation of Information Technology	Hierarchy of technology use ranges from basic to advanced inclusion in curriculums and instruction. Hierarchy is model-like and could be used to structure evaluations.	Because of the nature of information technology, multiple methods are inferred for its evaluation.	Insofar as there is a need to integrate information technology into the curriculum, the overall support system (macro level) and classroom (micro level) contexts come into play.	Taking full advantage of the computer for learning, teaching and accessing information is a simple yet intricate concept with many subtle forms of implementation.	The authors cite the need for understanding the complexity of information technology and the nature of its implementation and impact.
7. Complementary Approaches to Evaluating Technology in Science Education	Heavy emphasis on a model that stresses evaluating context as it impacts the ultimate success or failure of a program.	Complex, mixed methods were employed for both evaluations depicted in the text.	One of the evaluations examined contextual variables such as administrative support, communication, and others in detail.	Context and its many contributing variables are specified but in a general way.	The pressing urgency to better define and, in turn, come up with ways to measure the context construct are highlighted.
8. Evaluation of Science	A participatory type of underlying model	Reliance primarily upon observation	Allusion to the larger context but	Terms such as habitus, being-in/with,	Teaching, learning, and evaluating such

Table 2. Continued

Chapter	Use of Models	Methodology Focus	The Role of Context	Definition Considerations	Ideas about the Future
Teaching Performance Through Co-Teaching and Co-generative Dialoguing	a la illuminative evaluation (Parlett and Hamilton, 1976) or responsive evaluation (Stake, 1975).	and engagement within the teaching environment.	the intent is to creatively look at evaluating teachers within their specific classroom teaching situations.	coteaching, cogenerative dialoguing, metaloguing are used to portray the main concepts and methodology of the chapter.	processes are the collective responsibility of many involved parties and require their enhanced participation in that process.
9. Evaluating Science Inquiry: A Mixed Method Approach	A model is proposed for the inquiry process consisting of three parts—abilities, procedures, and classroom philosophy.	Numerous methods are used in the illustrative example in the chapter, including surveys, tests, observations, evaluations of work samples, etc. Multiple methods abound.	The classroom context is the main focus here and is most noticeable in the three part model in regard to the teacher's philosophy as it affects inquiry.	Multiple ways of defining the concept of process are examined with the explicit notion that it is complex and not unidimensional	New perspectives on the nature of inquiry processes will require mixing methods, and probably a more participatory approach to evaluation will emerge.
10. Distance Learning in Science Education: Practices and Evaluation	Not model driven but various modes of distance education are described which, in turn, would affect the evaluation of distance education programs.	The use of many methods for evaluation of distance education programs from the literature are given throughout the chapter.	While not dealing specifically with context, aspects of it are evident such as the adverse effects of distance education in terms of instructor time needed to answer e-mail, isolation, and problems in chat rooms.	The author defines distance education/learning but proceeds to illuminate the subtleties of the definition.	The growth of distance education and learning is inevitable, avenues for more research into this educational delivery system are suggested.

perspective of six years later, it is interesting, reassuring, and even refreshing to see such strong emphasis being placed by the authors of chapters on models/structures to direct their endeavors.

Turning to the **second heading** in Table 2, **mixed methodology**, nearly all of the chapters had either some or extensive writing on the concept, even though at times, the coverage may have been in a fashion that is not totally informed by the specialized literature on applications of mixed methods. The chapters seem to be saying that certainly mixed methods make good sense and should be used. To illustrate this, look at the Chapter on TIMSS (#2). The test results trigger a series of questions as to what might be affecting the performance of the United States relative to other countries. (That is, why don't we fare well in mathematics and science achievement as demonstrated on the tests and the numerous comparisons that have been made with other nations?) Answering a question of this type really requires the in-depth study of curriculum and instruction almost exclusively through qualitative methods, once deficiencies have been revealed by "harder methods", i.e., the tests.

Evaluating reform is another natural candidate for mixed methods. In the best of all possible worlds (an idealistic world *a la Candide*), perhaps we would create a series of quantitative performance indicators that could easily and routinely be monitored via database or bench-marking systems. Closer inspection tells us, however, that thinking this way may lead to unattainable goals compounded with frustration.

Reform takes place within diverse and subtle organizational contexts. It occurs when key individuals at all levels of an organization buy-in and commit over and above what would be normal amounts of work and emotional energy to the enterprise. Public acknowledging the need for reform and mandating it probably will promote some change, but not much along the lines of deep-rooted efforts that are sustainable in the long term. Furthermore, it would be difficult to define indicators (especially quantitative ones that capture subtle feelings and attitudes) and then to establish indicator systems that would ever begin to show the nature of what is really happening in a reform context? From our perspective, this is highly unlikely.

It is no surprise that, as pointed out in Chapter 3, the evaluations of reform will automatically include multiple quantitative and qualitative methods. The author of the chapter wisely describes many studies being carried out where such methods are being employed to characterize what reform is (what it looks like, what its parameters are, what measures of success are, the relationship of aspects of programs to measures of success, what types of leadership styles support change, etc.) and to isolate the most important factors propelling it to success. Chapter 9, as another example, not only has phraseology in its title referring to mixed methods, nearly the entire emphasis of its text is on the use of mixed methods to evaluate the process of inquiry in science education. Samples of instruments that were utilized for a specific mixed methods evaluation are also included in the chapter.

Without belaboring the issue, other chapters directly use or in other instances indirectly allude to multiple methods as the way to go. This is very appealing to us as co-editors and the older co-editor in particular. When he received his graduate education in research methods there was in reality or seemed to be (at least in his perception) only one preferred course of action to follow when evaluating/studying programs—quantitative methods and approaches. Collect numeric data and apply statistical analyses to that data—that was sole game to be played. At the present time, however, new sophistication has emerged with a readiness and an easy willingness to embrace multiple methods as the best means for obtaining in-depth understanding of programs especially as they interface and interact with the systems around them. The situations are too complex and any one method will be insufficient for the evaluation task.

It is satisfying now that the multiple methods paradigm is accepted and used by the chapter authors. Conversely, a word of caution is in order. Multiple methods affect the evaluations themselves in a number of regards. They require considerably more funds and time than a single method would and they tax the skills of any single evaluator. Even though most evaluators tend to be eclectic in terms of embracing multiple methods, it is a major stretch to believe that they will be equally adept and facile with radically different methods and up-to-date in knowledge about them. Generally, teams of evaluation personnel will have to be assembled to implement multiple methods evaluations and thus the costs of the evaluations will not only shoot up, the evaluations will take more time to implement and the results will be harder to analyze and interpret.

Another issue here revolves around the meaning of the term "mixed" or "multiple" methods. In this discussion they have been used interchangeably with about the same connotation. In truth, there are subtle considerations that have to be made when thinking about applying more than one method to an evaluation. Greene, Caracelli, and Graham (1989) in a classic paper have noted that there are five ways (e.g., triangulation, complementarity, etc.) in which methods are used in a joint manner.

More recently, Altschuld and Witkin (2000) described (in needs assessment contexts) what they called "within method" and "between methods" variations of mixed methodology. In a within method variation, one method has three or four different versions of it being employed as in three or four different yet similar questionnaires intended for different audiences and different audience response sets. A between methods variation is where varied methods (surveys, interviews, analyses of records) are used to gather data. Another consideration is that different methods can be used for samples of individuals involved in a specific program thus further complicating the evaluation. Whatever way we choose to name or characterize the multiple methods picture, it is a much deeper construct when examined under the fine lens of a high powered microscope.

Before leaving multiple methods one further concern warrants mention. Altschuld and Witkin (2000) reviewed five needs assessment studies that were based on the use of multiple methods. In all but one, the results did not agree although to some level having results from more than one method was helpful in understanding the problematic nature of the area under investigation and issues related to it. Returning to what we have learned from the chapter authors, there seems to have been a fairly pronounced emphasis on multiple/mixed methods which is commendable, but at the same time comprehending the full implications of such use is itself a complex phenomenon and one to which we will have to attend to more in the future.

Shifting to the **third** major column in the table, **Context**, appears to be a critically important theme for evaluation in science and technology education. A bias might be evident here. The two of us have written about how context affects the ultimate success or failure of programs. We have even proposed a model in science education to evaluate features of context as they affect programs. Admittedly there is a predisposition to thinking about the context in evaluative studies. Dealing with it could be merely a reflection of self interest, in the sense of a self-fulfilling prophecy. This could have happened but we don't think that it did.

Context, as a construct, is notable in a number of chapters. The one that looks at TIMSS (#2) goes way beyond the tests and the information they provide. If test data were the only source then there would be limited information upon which to formulate policies about science instruction in U.S. schools. The authors found it mandatory to seek an understanding of how curriculums and instructional approaches in the U.S. are similar or dissimilar with those in other nations. Why is there such a sharp decline in achievement for our 8th graders in comparison to their 4th grade scores. To what in school policies and efforts can these observations be attributed? Without exploring the context, limited utilitarian information is gained from the tests as a stand-alone approach to evaluation.

Previously we noted the model or framework of Chapter 3 which the author provides as a means for structuring the evaluation of reform efforts. Now we would like to revisit that model and proclaim it as also having context as its main feature—its central element. There is clear recognition that without a supportive context, reform efforts and major new initiatives will tend to fail. This is not an innovative concept nor one that has not been raised before or investigated in many ways (see for example, Bhola, 1965; Hutchinson & Huberman, 1993). But it is one that is resurfacing in a more sophisticated manner especially as we move into an era of rapid educational and technological change.

The concern is apparent in other chapters with reference to various levels of context. In Chapter 8, the authors are thinking of the micro-context of an individual classroom and how in the evaluation of teaching, that micro-context is paramount, it is where the focus must be. In fact, they recommend a completely different way of evaluating pre-service teachers that is micro-context dependent. Chapters 4 and 5 are analogous in this regard by moving from the macro-level of context (the overall system level) to the classroom where ultimately all changes really take place. It is almost like there is a dynamic tension between the system (macro) and the micro levels akin to what Welch (1974) described more than a quarter of a century ago. Chapter 8 goes further by addressing some of the tension between these levels and how it affects the classroom situation.

Context is embedded in one form or another in nearly all of the chapters. Aside from sorting out levels of context, a problem arises in trying to define what context is and trying to make it operational within educational settings comprised of so many actors and features. Considering just the drivers of reform, how are they measured? What are the specific variable manifestations of them? What are the quality (reliability and validity) of measures of them? What are the best methodological strategies to use when assessing the degree of their presence or absence? When should they be assessed—very early in the life of an effort or maybe later if they are the types of variables that change and begin to appear over time? Can the individual effects and contributions of such variables within complex educational milieus be ascertained with reasonable certainty?

The difficulties of measuring variables is directly illustrated via the problem that Cullen and her colleagues (1997) encountered in an evaluative study of science education reform. They visited four sites (2 rural and 2 urban) that had been cross-nominated as exemplars of educational reform in the U.S. To conduct their case studies, the evaluators prepared semi-structured interview protocols for educational personnel (teachers, system administrators, curriculum supervisors, etc.) at each site. When the protocols were first implemented, it was immediately apparent that teachers were not resonating with or understanding the questions because they had been developed from primarily a system rather than a teacher and classroom perspective or point of view. The investigators were forced to change their prompts to reflect the environment of teachers and thereby to encourage them to respond more than they had done so in the initial round of interviews. This example goes well past showing methodological complexity, it also provides a look at how invaluable it is to consider macro-micro issues when evaluating reform oriented programs. How reform was thought about at the macro-level, the system level, became quite a different thing at the micro-level, the level of teachers and students.

While there are a variety of tough questions to answer about context, the idea of investigating, exploring, and evaluating the construct, is fascinating. It seems that the chapter authors are opening the door to a future agenda for research. The quality of the innovation or the new effort will, of course, have an impact on its eventual adoption and use, but in schools the context may just simply overwhelm many good ideas. The context may not have enough change orientation and by that very nature (that atmosphere) almost snuffs out the flame of a new direction. Indeed, given the omnipresence of context and the importance of its influence on change and reform, we may have to put more of our resources into its evaluation than is currently practiced. Furthermore we will have to do a better job of defining the construct and creating suitable measures of it for evaluation.

Indeed, how we define and deal with key constructs, terms, and variables is absolutely vital to moving forward in the evaluation of science and technology education programs and initiatives. **Definition Considerations**, the fourth substantive column heading, contains brief illustrations of the difficulty of the issues confronting the authors in this regard. For example, what is meant by policy and policy formation? What are the implications of policy and what role does it play in the United States as compared to other countries? What would the policy look like, what form should it take? Who is and who should be part of policy debate and deliberations? How does policy formulated at the highest levels of government become a reality in an educational system characterized by local control? How does policy get translated into action? Does that translation (or perhaps transformation) have fidelity in terms of what was intended by the policy? How are policy initiatives and, in turn, new directions and programs funded? Does high level policy formation imply a federal or state level system of education? If this were to occur, what are the implications for evaluation? (On this latter point, see Cohen, 1970).

Analogously, problems in definitions are apparent throughout the chapters. Another specific example is found in Chapter 8 where a sizeable portion of the text is devoted solely to analyzing the possible ways in which inquiry and process skills are defined in the literature. The authors of Chapter 6 note right at the beginning that they had been given a very daunting (maddening in some respects) task as their assignment. What is information technology in general and what might it look like in science education, in specific? Distance education as another concept can take so many forms and shapes that it virtually becomes multidimensional as opposed to uni-dimensional. Many of the terms in the chapters could be thought of in this way.

In Chapter 8, the evaluator as a co-teacher and a co-participant in the classroom is a powerful and thought-provoking concept. The positive benefits of the evaluator's (if one can now be called that) presence in the classroom as a co-teacher and co-participant are unmistakable. One cannot deny the value to be gained from this point of view in training and developing the future cadre of science teachers especially in the urban setting as contrasted with the traditional perspective of the evaluator being seen as a person who renders judgments. The chapter demonstrates the contrast well.

But what is being proposed is a subtle concept with somewhat amorphous boundaries. Is it possible that the co-teacher, the co-participant, and the co-generative dialoguing could be a little too close to the situation and so involved in its intricacies that a portion of objectivity gets lost in the process? Is that loss good or bad? What is the role of objectivity here or perhaps our whole pattern of thoughts about what constitutes reality and meaningfulness? Are our current paradigms in need of radical alteration?

The evaluator is now a teammate in the drama, a fellow cast member in an active, supportive role in a live play. The demands of this stance on the development of reflective teachers cannot be minimized in regard to the press they will place on teacher educators and teacher education institutions. Those demands will be high and they will force the reexamination of the underlying assumptions of teacher training.

Context as a variable to which all of the authors refer in some way or another as well as to varying extent is obvious. Returning to an earlier point, closer inspection, however, reveals that context has many different meanings to these individuals. Some chapters are dealing with either the macro (school system) context or to what might be termed the super macro (state and federal level) context. Others seem to be focused on the more micro (school or classroom) level. Others allude to the interface between these levels as being of major importance and that the micro level has to take a more active stance than it has in the past in reform and accountability efforts. In other words, some are suggesting a grass roots voice must be heard in the policy debate, particularly as related to accountability and testing.

In no way are these observations about the complexity of the term intended to reflect the efforts and hard work of the chapter authors. Rather they direct us to much different conclusions. First, concepts and terms are not easy to work with and they do and will continue to tax our ability to generate definitions. Second, it must be recognized that simple definitions are just that—simple, general rubrics that at best convey the sense of an idea, but begin to almost mutate, erode, and to use a stronger word, essentially deteriorate as you move beyond the simplistic stance. The proof is in the pudding and virtually most of the terms are filled with nuances that have effects on the evaluations of programs and projects. Finally, in some cases we noticed that different terminology was assigned to what seemed to be identical concepts. This is best illustrated in Chapters 2 and 4 where curriculum alignment and coherence integration, respectively, were used for what seemed to be the same concept or ones that are similar in nature.

Some common and accepted definitions would most likely have value for the conduct of evaluations and for comparing evaluation studies in science and technology education. Altschuld and Witkin (2000) found an analogous situation in the field of needs assessment and as a result they created a glossary of definitions that was viewed as having utility for the work of needs assessors. Perhaps a related effort would be of value for the substantive foci of this book.

WHAT DOES THE FUTURE HOLD IN STORE FOR THE EVALUATION OF SCIENCE AND TECHNOLOGY EDUCATION?

We are exercising a prerogative afforded to co-editors by substantially departing from the prior format. The thoughts and ideas in this section are not totally dependent on the writing of chapter authors. Assuredly, although some of our conclusions are attributable to them, they more emanate from personal perceptions based on forty years of collective work in evaluation. What stands out to us, what do we think the future holds?

FIRST AND FOREMOST, THE FUTURE IS EXCITING AS IS APPARENT IN THE CHAPTERS

We did initial forays into the literature of the evaluation of science education around 10 years ago. There was no TIMSS at that time and reform movements were in their infancy or perhaps in a nascent state as was the notion of information technology and its potential for instruction. The literature reviews we conducted then would be best characterized as disappointing and that would be a serious overstatement. The chapters show a different and emerging picture that represents vast change over the decade. The needs for evaluation, research into evaluation, and research into the nature of science and technology education programs seem unlimited and huge.

It is easy to demonstrate this point. Consider the topic of reform in schools. What constitutes successful leadership of reform and what kinds of evaluative information would be most useful to leaders in this regard? What combination of drivers have to be present and what is their relative weight in terms of promoting reform? Are some set of factors or the absence of some set always associated with success or failure of reform efforts?

Think of distance education, how we will be evaluating the quality of instruction and its impact on students, teachers, programs and institutions? What types of methods and approaches will we have to create or modify to fit with a vastly different instructional delivery mode involving technology?

Or looking at even another area, testing and accountability systems, how do teachers perceive them and what changes do they foster in classrooms? If test scores rise, is that result caused by solid and fundamental changes in curriculums, content, and instructional practices or could it, in some cases, be attributed to teaching to the test? In other words, are there negative, deleterious impacts of systematic testing and accountability structures that actually work or mitigate against real and meaningful change? What do our logic maps tell us about such circumstances? How do we progress educationally? How should we as evaluators facilitate and take part in the deliberations that lead to progress?

One could take any of the chapters and generate many other questions for research and evaluation. The authors have simply, but in a thoughtful way opened the door to new avenues of inquiry and by so doing encouraged researchers, evaluators, supervisors, policy-makers and graduate students to pursue them or questions of their own choice. We are heartened by this state of excitement in evaluation. What a difference from 10 years ago! Evaluation is alive, well, and kicking in science and technology education.

SECOND, WHAT ARE THE IMPLICATIONS FOR TRAINING THE NEXT GENERATION OF EVALUATORS?

While it was not the charge of the authors, their writings suggest that we must look more closely at the training of individuals who conduct evaluations of science and technology education programs. In 1994, Altschuld et al. published a directory of evaluation training programs in the U.S., Canada, and Australia. Forty-nine programs were identified at that time, but many were seriously limited with only about 15 programs having 3 or more courses. The 1994 study, which is currently being updated (Engle and

Altschuld, personal communication, May 29, 2001), revealed that most of the major programs (size wise) dealt with generalizable types of training, e.g., models of evaluation; exposure to the main principles of conducting evaluations in any field.

The assumptions upon which most programs operate, however, have been called into question. Indeed, in the late 1990s The National Science Foundation funded four university-based training programs focused specifically on the development of evaluators with knowledge of math and science education. The Foundation also underwrote several other evaluation training endeavors. (See Frechtling and Sanderoff, 2000, for a discussion of these initiatives.) One key concern of NSF is that evaluators should know more about the substance of programs they are evaluating. How well has this training worked and the validity of its underlying premise are currently under scrutiny since NSF funding for the four programs will be ending in 2001.

More recently the Foundation has been looking at the cultural competencies needed by math and science evaluators who often find themselves involved in programs located in increasingly diverse educational and community environments, especially the urban school systems. What should evaluators know and understand to be successful in very specialized situations? And relatedly, how do we find and encourage the next generation of evaluators particularly those from under represented groups? (See Johnson, 2001.)

So it could be inferred from the chapters that one overriding issue for the future of evaluation of science and technology programs is that of training evaluators. What blend of technical competence, knowledge of the field of evaluation, knowledge of the discipline and/or area to be evaluated, and personal skills are necessary for developing evaluation talent or for building teams of evaluators? How will we recruit people into evaluation? What will be required to keep the pool of evaluators updated in regard to the rapidly changing nature of the field? What is(are) the best mechanism(s) for the delivery of evaluation training. Who should provide training?

What should be the balance between theoretical and applied work in training? That is perhaps a more subtle and difficult concern to which we must additionally attend. Evaluation has been referred to as a "near field" (Worthen, 1994) or a "transdiscipline" (Scriven, 1994) and clearly the views offered by the chapters seem to support these perceptions. Therefore, can programs be just "applied" if we are to maintain the idea of a field of evaluation? Evaluators have a professional responsibility to develop training programs that span the practical and theoretical demands that undoubtedly will be placed upon them.

THIRD, FOR WHAT ROLES ARE SCIENCE AND TECHNOLOGY EDUCATION EVALUATORS BEING TRAINED?

We might ask, as an outgrowth of the prior question, for what roles are evaluators being trained? Over 30 years ago Stake (1967) proposed that evaluators served in three distinct capacities—describers of programs and projects, facilitators of decision-making, and as active participants in the decision-making process (that is, as one set of decision-makers or judges). His views are still potent this many years later.

Certainly, we are describers and we must continue to provide cogent, detailed, and meaningful portrayals of context, even though the task is arduous and filled with pitfalls. The programs and projects included in the chapters are indeed complex and will press our ability not to just describe them, but to describe them in ways that capture the subtle nuances of important features and critical human interactions.

It is the other two roles of Stake that are much more problematic and complicated to address. Consistently, many of the authors were dealing with the translation of evaluation findings into recommendations for change especially in the policy arena with its associated politics. Education reform is really rooted in the classroom, the teacher, and the approach to instruction and, therefore, the voice of teachers must become more instrumental in the accountability/testing debate. And somehow, through us, that voice has to become more explicit and well articulated.

What stance do we as collectors of information play in this game? How do we or how should we facilitate the nature of decision-making? Do we know enough about transforming our findings into policies that will help in creating reform and enhanced educational opportunities? Whether we like it or not such roles may be forcefully thrust on us.

Consider the case of the effects of TIMSS and the subsequent investigations of curriculum and curriculum delivery systems that it engendered. It is not surprising, most likely as a direct result of the findings, that the American Association for the Advancement of Science (AAAS) called for streamlining the science curriculum in American Schools (Henry, May, 2001). Streamlining includes a reduction in the number of topics covered, removing unnecessary details from topics, and limiting technical vocabulary. One criticism of this change in policy has centered on the potential of creating a sort of second class citizenry by essentially "watering down" the quality of science taught in the country. And of course, there has been a response to that criticism.

So the question remains what roles do we play as describers, judges, and facilitators of decision-making? What does it mean to involve multiple educational constituencies in decision-making? How should we involve

them? How do we facilitate the making of decisions? Do we fully understand the negative and positive consequences of our findings and the new
policies to which they might lead? Are we training evaluators for such contingencies and assuming that the answer is generally "no", how might we
go about doing so in science and technology education?

FOURTH, NEW WAYS OF THINKING ABOUT
EVALUATION ARE EMERGING

To reiterate something that has been mentioned before and to make sure
it does not get lost in the discussion, the chapters are rich sources about a
number of new evaluation topics or ways to think about evaluation. Some
of them are: the use of mixed methodology; the nature of how to evaluate
and investigate reform efforts; the stance and role of the evaluator in pre-
service teacher education; the dimensions of information technology and
distance education and what it will take to evaluate these entities; and the
numerous considerations raised with regard to standardized testing
programs and accountability systems as well as the role of and the need
for alternative testing approaches. Before encouraging you to explore
the chapters let us offer a final word about some things not covered in the
book.

Any book has limitations in what it is able to cover and how it is
framed by the authors, editors in this case, and this one is no exception to
that rule. Due to space and time considerations, evaluation of non-formal
science education and comparisons of post secondary vs lower level evaluations were a few of the themes not included in the prospectus. Either use
the chapters as a springboard for expanding the repertoire of ideas proposed by the authors or move into new areas with which evaluators
should be familiar. **In either case we simply ask that you please accept the
challenge.**

REFERENCES

Altschuld, J. W., and Witkin, B. R. (2000). *From needs assessment to action: Transforming needs
    into solution strategies.* Thousand Oaks, CA: Sage Publications.
Altschuld, J. W., and Kumar, D. D. (1995). Program evaluation in science education: The model
    perspective. *New Directions for Program Evaluation,* 65:5–17.
Altschuld, J. W., Engle, M., Cullen. C., Kim, I., and Macce, B. R. (1994). The 1994 directory of
    the evaluation training programs. In Altschuld, J. W., and Engle, M. (eds.), Preparation
    of professional evaluators: Issues, perspectives, and programs. New Directions for
    Program Evaluation, No. 62, pp. 71–94.

Bhola, H. S. (1965). *The configurational theory of innovation diffusion.* Unpublished dissertation, Columbus, OH: The Ohio State University.

Chen, H. (1990). *Theory-driven evaluations.* Newbury Park, CA: Sage.

Cohen, D. K. (1970). Politics and research: Evaluation of social action programs in education. *Review of Educational Research,* 40(2):213–238.

Crichton, M. (1999). *Timeline.* New York: Ballantine Books.

Cullen, C., Denning, R., Haury, D., Herrera, T., Klapper, M., Lysaght, R., and Timko, G. (1997). Case studies: Teachers' perspectives on reform and sources of information. (Technical report from The Eisenhower National Clearinghouse for Mathematics and Science Education). Columbus: The Ohio State University.

Frechtling, J., and Sanderoff, H. (2000, August). Proceedings from the NSF workshop on training mathematics and science evaluators. Rockville, MD: The Westat Corporation.

Greene, J. C., Caracelli, V. J., and Graham, W. F. (1989). Toward a conceptual framework for mixed-method evaluation designs. *Educational Evaluation and Policy Analysis,* 11(3):255–274.

Henry, T. (2001, May 24). Group dissects science literacy. *USA Today,* 11D.

Hutchinson, J., and Huberman, M. (1993, May). Knowledge dissemination and use in science and mathematics education: A literature review. A report prepared for the Directorate of Education and Human Resources, Division of Research, Evaluation, and Dissemination, National Science Foundation under the Order number CB264X-00-0.

Johnson, E. (Ed.) (2000, June 1–2). *The cultural context of educational evaluation: The role of minority evaluation professionals.* Workshop proceedings, Directorate for Education and Human Resources, Division of Research, Evaluation, and Communication, The National Science Foundation, Arlington, VA.

Kumar, D., and Chubin, D. E. (Eds.) (2000). *Science, technology, and society: A sourcebook on research and practice.* New York: Kluwer Academic/Plenum Publishers.

Parlett, M., and Hamilton, D. (1976). Evaluation as illumination: A new approach to the study of innovatory programs. In G. V. Glass (Ed.), *Evaluation Studies Review Annual, 1.*

Rae, A. I. M. (1994). *Quantum physics: Illusion or reality?* Cambridge, NY: Cambridge University Press.

Scriven, M. (1994). Evaluation as a discipline. *Studies in Educational Evaluation,* 20:147–166.

Stake, R. (1967). The countenance of educational evaluation. *Teachers College Record,* 68:523–540.

Stake, R. E. (1975). Program evaluation, particularly responsive evaluation (Occasional Paper No. 5). Kalamazoo: Western Michigan University Evaluation Center.

Welch, W. W. (1974). The process of evaluation. *Journal of Research in Science Teaching,* 11(3):175–184.

Wholey, J. S. (1987). Evaluability assessment: Developing program theory. In L. Bickman (Ed.), Using program theory in evaluation. *New Directions in Program Evaluation,* 33:77–92.

Worthen, B. R., Sanders, J. R., and Fitzpatrick, J. L. (1997). *Program evaluation: Alternative approaches and guidelines.* New York: Longman Publishers.

Worthen, B. R. (1994). Is evaluation a mature profession that warrants the preparation of professional evaluators? In J. W. Altschuld and M. Engle (Eds.), The Preparation of Professional Evaluators: Issues, Perspectives, and Programs, *New Directions for Program Evaluation,* 62:3–16.

CHAPTER 2

What Role Should TIMSS Play in the Evaluation of U.S. Science Education?

William H. Schmidt and HsingChi A. Wang

INTRODUCTION

The Third International Mathematics and Science Study (TIMSS), the most extensive and far-reaching cross-national comparative study of mathematics and science education ever attempted, has influenced major changes in the policy and practices of mathematics and science education in many of the participating countries. The results of TIMSS sparked vigorous discussions within the United States, especially. Some of the excitement is simply due to the fortuitous timing of the results, released at a moment of tremendous public interest in education and when some states were moving towards standards-based education. Since then TIMSS has come to

William Schmidt, Department of Counseling, Educational Psychology and Special Education, College of Education, Michigan State University, 463 Erickson Hall, East Lansing, MI 48824-1034.   HsingChi Wang, U.S. National Research Center, Michigan State University, East Lansing, MI 48824.

Evaluation of Science and Technology Education at the Dawn of a New Millennium, edited by James W. Altschuld and David D. Kumar, Kluwer Academic / Plenum Publishers, New York, 2002.

be regarded as an authoritative source of information about policy and practice in educational reform.

TIMSS has compared the official content standards, textbooks, and teacher practices and student achievement of almost 50 countries. Hundreds of content standards and textbooks were analyzed, representing some 30 languages. Thousands of teachers, principals, and other experts responded to questionnaires, while over a half million children worldwide were tested in mathematics and science. The tests were administered to 9-year-olds, 13-year-olds, and students in the last year of secondary school. For most countries, this meant that the testing was done at grades three and four, seven and eight, and the end of secondary school[1].

Although TIMSS seems likely to influence policy and practice in science education in the United States this should not cloud the more fundamental question of what role it should play in the *evaluation* of U.S. science education. It is this question to which this chapter is addressed.

On the face of it, **TIMSS should play no role**—it should not be used in the evaluation of particular science programs nor should it be used to evaluate general science reform efforts. The study was never designed to address such issues, and to use it in this way would be inappropriate. This is especially true given the "splintered" nature of U.S. science education (Schmidt, McKnight, & Raizen, 1997).

In the United States each of the approximately 16,000 local school districts have their own science education standards, and all the states except Iowa have state standards; when one considers this together with the American Association for the Advancement of Science's (AAAS) *Benchmarks* and the National Academy of Science's (NAS) *National Science Education Standards*, one finds a plethora of visions without any coherence to bring them together. One would think our model for science education was the Tower of Babel. Under these circumstances where there is no national program, to use TIMSS as an evaluation of programs or reform efforts would be sheer folly because nationally representative samples of students such as those used in TIMSS would not provide representative data for any particular program or reform effort.

TIMSS can and should play a role, however, if the focus of the evaluation is on the **U.S. system of education as a whole** and not on particular programs or reform efforts. Included in such an evaluation would be legitimate questions such as how U.S. science education is organized (including the very nature of the splintering referred to above) and what the general characteristics of the curriculum are especially in terms of content

---

[1] TIMSS-Repeat (TIMSS-R) was recently completed in 1999 for 38 countries and it too examined some of the same issues but focused only on eighth grade.

coverage. Since TIMSS has a nationally representative sample of U.S. students and comparable samples from other countries, then for this type of evaluation—the system level—TIMSS would have an appropriate role. But that role would be one of generating hypotheses that warrant further examination of U.S. science education.

REFORMS IN U.S. SCIENCE EDUCATION

During the past four decades, science educators have been actively considering ways to address the problem of scientific illiteracy. Wang's (1998) review of the historical efforts in U.S. science education reform groups them into three major periods: 1) the golden age of science education: Post-Sputnik reaction, 2) science education for an enlightened citizenry, and 3) standards-based science education.

The launch of *Sputnik I* also launched the "Golden Age of Science Education" (Kyle, 1991) in the United States with innovative and spectacular changes in American science education during the mid 1960s and early 1970s. So that future scientists could compete with the pace of scientific research in the Soviet Union, the public demanded more rigorous science education; as a result "alphabet-soup" science curricula were much improved. Unfortunately, this wave of reform failed to integrate the curriculum with other sectors of education—schools, curriculum developers and distributors, higher education institutes, and so forth; thus the materials (developed mostly by scientists and higher education faculty) became a collection of isolated projects that teachers could not use to their best advantage.

The call for an enlightened citizenry rather than an "educated elite" for society (DeBoer, 1991) initiated a second attempt to reform science education starting in the early 1970s. There was a concern that the perception of science as rigorous might discourage potentially capable students from taking science classes, so in the interest of educating the general public and increasing scientific literacy, reforms in the name of "science for all" called for curricula that would be relevant and appealing to every student.

The reform movement pushed for constructing a humanistic, value-oriented curriculum that would portray a wide range of personal, societal, and environmental concerns (National Science Teachers Association, 1982). Research based on studies of humanistic science programs such as Mascolo (1969) and Welch (1973) indicated that these approaches had a positive impact on students' inquiry ability and science understanding. In addition, a study by Quattropani (1977) showed that the changes in the science curriculum significantly increased science enrollment. Still, teaching for understanding requires teachers to have knowledge in both inquiry pedagogy and subject matter. Unfortunately, what also has been found is that the limited content-knowledge background of teachers continued to be the factor deterring the majority of teachers, who implemented the change, to buy into and practice the philosophy of teaching for understanding.

The traditional lecture format offering fragmented pieces of factual information seemed more comfortable to teachers and was perceived as a more efficient method of delivering the "hard sciences" than humanistic or context-based science teaching. The humanistic approach to science education was almost thrown out when the "back to basics" movement arrived during the 1980s. As observed, these two camps—"teaching hard sciences" and "teaching for understanding"—seem to have begun their wrestling match fairly early in the history of American science education.

Stepping back from the debate between content versus inquiry, the National Commission on Excellence in Education (NCEE) released *A Nation at Risk* in 1983, a major report that ignited the movement to march into the standards-based science education reform era. Educators and all stakeholders enthusiastically engaged in establishing educational standards for science curricula to enhance all students' scientific literacy. Two major professional organizations—AAAS and NAS/NRC—proposed standards for what students ought to learn and be able to do in science (AAAS, 1993; National Research Council (NRC), 1996). In short, the science standards defined what constitutes scientific literacy. The *Benchmarks for Science Literacy* (AAAS, 1993) first defined the content standards for science education. The NAS/NRC then elaborated on AAAS's content standards with standards for designing assessment, professional development, teaching, programs, and systems in science education.

This new wave of standards-based reform in science education finally recognized that, for American education, *the curriculum* is what needed to be seriously examined and reconstructed in order for a sensible education reform to take place. To put it simply, our science education consists of fragmented visions. There is a pressing need to focus on what students need to know and be able to do at the end of their formal education.

Unless the curriculum is focused and coherent, policy and practice will continue to produce fragmented projects that fail to coalesce and teachers will continue to receive training that ill prepares them to teach essential science. Without coherence at either of these levels, it remains difficult to evaluate the effectiveness of science instructional materials that were produced to appeal to all kinds of visions by districts and states. Quality instructional materials are a critical element to quality science teaching as described by one of the six Science Teaching Standards stated in the *National Science Education Standards* (NRC, 1996):

Effective science teaching depends on the availability and organization of
materials, equipment, media, and technology. . . . Teachers must be given the
resources and authority to select the most appropriate materials and to make
decisions about when, where, and how to make them accessible. (NRC, 1996,
pp. 44–45)

The unfocused nature of American science education fails to inform
the system as to how to adequately prepare teachers to teach science. It
fails to reliably assess students' knowledge of science, and it cannot provide
effective criteria to evaluate instructional materials to assist quality science
instruction.

WHAT ROLE SHOULD TIMSS PLAY?

TIMSS was designed to collect data pertaining to the entire education
system of a country. The underlying conception for the study is based on an
"opportunity to learn" model that can be found in Schmidt and McKnight
(1995). This model suggests that the relationship between curricular experiences (educational opportunity) and learning defines the essence of
schooling. Students' experiences are ultimately a matter of how each individual integrates what comes in from the outside with what goes on in the
inside. Teachers may help create and shape experiences, but students' actual
experiences result from integrating those opportunities with their own
engagement in those activities. The combination of opportunity and engagement determines the actual experiences flowing from encounters with
potential experiences. Education systems plan the creation of educational
opportunities. For instance, teachers shape learning experiences. Students
then reshape those opportunities; they act as the final arbiters in creating
their own experiences. It is out of those actual experiences that learning
occurs.

The "opportunity to learn" model that undergirds TIMSS considers
curriculum only in its role of distributing potential experiences. This conception of educational opportunity recognizes that education policymakers
greatly influence and shape the potential educational experiences of students. This is the central role of a nation's education policymaking body. As
such, this model suggests that the emphasis of the evaluation to which
TIMSS can contribute must be the education system itself, i.e., toward influencing the nation's policymaking bodies.

Given this conception and model, the measurement techniques and
instruments developed as a part of TIMSS focus on assessing the education
system and the way in which it provides educational opportunities to students. The role that TIMSS can play in the evaluation of science education

must be aimed at this level. In other words, TIMSS is designed to examine the effectiveness of an education system in creating and providing educational opportunities for students. Played out in this context, TIMSS does have a role to perform in the evaluation of science education in the United States.

One of the inherent difficulties is that, unlike most other countries, the United States has no national vision for science education. In fact the policy making body for the U.S. is many such bodies. What then is the system that TIMSS can help evaluate? Can we view science education nationwide as a single system? That is the heart of the problem in using TIMSS for such a purpose.

For TIMSS we defined the U.S. education system as an aggregate of the state systems. For the curriculum phase of TIMSS we examined the content standards of a random sample of states, and, for the purposes of the achievement and survey questionnaire data, nationally representative samples were chosen to represent the United States as a whole. The generalizations from TIMSS must be applied only to this composite which we term the U.S. education system.

Thus using TIMSS as an evaluation instrument is not specific to any particular district or state, or to any particular program tried within any district or state. The TIMSS evaluation in this context is applicable to a composite of the U.S. education system that in some sense represents some 16,000 local school districts and the 50 state school systems.

One difficulty with this conception is that there is a great deal of variability within the U.S. system. This variability is a characteristic of the system and as such is part of what can be examined in such an evaluation. In other areas of economic, social and public policy the U.S. has no difficulty in conceiving of itself as a single country. Why should this be different for education? From an empirical point of view, the typical TIMSS results suggest greater commonalties across states and districts on many aspects of education but especially the curriculum than would be thought to be the case. This is especially evident when variations at the state and district levels are examined in light of the variations across countries. The latter are larger for most aspects of education than the former. In other words, variations across countries are larger than variations across districts or states within the United States. This reinforces the argument that it is reasonable to use the U.S. composite since it likely represents what science education looks like.

The implication of all of this and the answer to the question posed in this section—what role should TIMSS play in evaluating U.S. science education—is to gauge the broad features of U.S. science education as contrasted with those similar features in other countries and to examine their

Table 1. Percentage of U.S. Seventh and Eighth Grade Students Taking Each
Type of Science Course

	Seventh Grade	Eighth Grade
General Science	29.6 (5.2)	28.4 (3.0)
Physical Science	4.0 (2.0)	29.2 (4.8)
Life Science	56.7 (5.3)	7.7 (2.7)
Earth Science	9.7 (3.7)	34.7 (4.2)

Note: The numbers in parentheses are standard errors.

general effectiveness for the nation as a whole. Three such broad features
of the system can be examined through TIMSS: the organization of science
curriculum in the United States, the coherence of U.S. science education,
and its rigor. The next three sections address issues related to these three
features, respectively.

THE ORGANIZATION OF THE SCIENCE CURRICULUM

Science education in the U.S. is structured somewhat uniquely when compared with other countries. Middle school science in the United States is
organized in what some have termed as the "layer cake" approach. Three
areas of science are considered appropriate for middle school students:
earth science, physical science, and life science. Each of the three courses is
rotated over grades six, seven, and eight. Which subject matter is covered
in which grade depends upon local district or state policies. In other words,
there is no uniform approach across the United States as to what seventh-
grade science entails. On the other hand, data from the TIMSS study summarized in Table 1 show that the predominant course at the seventh grade
is life science, while there is a fairly even split between general science, physical science, and earth science at the eighth grade.

Two key features emerge with respect to the structure of science offerings in the middle schools. First, the sciences are separated into distinct
courses and covered in separate years. Second, although there is a general
pattern as noted in Table 1, the sequence of particular courses over the
middle school grades is not uniform across districts or states. In other words,
two districts, very close to each other geographically, might have different
sequences for these three courses.

On both of these points the U.S. approach to the organization of
school science is quite different from much of the rest of the world. In other
countries science is typically not segregated into separate disciplines and

then taught separately in different years. In countries such as Japan, a single science course is taught at the eighth grade but the topics included in that course come from several different disciplines, such as earth science, biology, and chemistry. But eighth grade science in many European countries includes from two to four separate courses, each covering one of the disciplines of science. In short, the first structural feature of American middle school science education, which separates the sciences by year, is not a common practice in other countries.

As to the second structural feature of U.S. middle school science, that is, the variation in subject matter across districts among the three grades differs completely from the practice of most of the TIMSS world. Whatever course structure, or whatever organizing principle they choose for their science instruction, it is the same across the country. Thus, in the United States the sciences are separated from each other instructionally across the three middle school grades while the sequence itself varies within grades across different parts of the country.

Given the relatively poor performance of U.S. students in science at the eighth grade, especially when contrasted to performance at fourth grade, one would need to at least question the way those courses are structured and offered in U.S. middle schools. This can only serve as an hypothesis; since the TIMSS study does not allow for the causal interpretation that would suggest that because of the course structure in the United States, the students do relatively poorly compared to those in other countries. Given the relatively poor performance of the U.S. students and the fact that there is a decline in the relative standing of science scores in the United States from fourth to eighth grade when compared to the rest of the world, such structural features of the organization of school science surely become important possible correlates of that declining U.S. performance. And perhaps even more importantly become important characteristics worthy of further study.

The issue of course structure involves the organization of not only the middle school curriculum but also the high school curriculum. Even though the data here are less extensive because of the nature of the TIMSS data collection, the available information suggests that the way courses are organized at the high school level differs appreciably between the United States and other countries. The U.S. high school science curriculum continues the same layer cake approach by offering different science courses in succeeding years—most traditionally, but not always, in the order of biology, then chemistry, and then physics.

In other countries, students are once again taking several courses during the same year. Material pertaining to a particular discipline is not isolated into a single course but spread over several years of instruction. In the Scandinavian countries the physics that is taught to science majors often entails the same material as covered in U.S. physics classes, but it is organized over two or three years. From an instructional point of view, one has to question the wisdom of trying to teach "all" of physics in a single year, when other countries may spread those same instructional topics over two or more school years.

A third structural feature of U.S. science education is tracking. In middle school science course tracking is not nearly as common as it is in the area of mathematics, but an interesting result from TIMSS shows that one of the perhaps unintended side effects of tracking for mathematics instruction is that this carries over into science instruction. Data from TIMSS indicate that in the United States students who are tracked into the higher levels of mathematics such as algebra are often, by schedule perhaps or by intent, tracked into specific sections of science courses. Even if not overtly designed to exclude some students, this likely produces differences in science instruction related to the perceived ability of the students. In other local systems, explicit tracking is practiced. It is clear that, even if not by explicit design, the structural feature of the way middle school science is organized can have an adverse impact on the stated goal of providing good science instruction for all U.S. students.

We have attempted in this section to show how TIMSS can be used in an evaluative capacity to address broad-based issues of how American science education is organized. The very organization of the U.S. curriculum both in the middle grades and the high schools appears from an international point of view to be somewhat unique. Although TIMSS was not designed to directly evaluate the impact of such structural features on science achievement, by drawing contrasts with the rest of the world in our relative achievement and on these structural features it does at least raise a question: which such structural features may be correlated to the corresponding differences in learning. Such hypotheses warrant further research.

It is in this way that TIMSS can be used to help evaluate U.S. science education. Perhaps the best description is that TIMSS serves in an evaluative capacity as an hypothesis-generating study. Clearly, before U.S. educators would be likely to conclude that the structure of our courses in science is the cause—or one of the causes—of our relatively low international standing in achievement, more data specific to this assumption would need to be collected.

The voice of our academic selves offers a side comment intended to be provocative to the science education community. Given that many policies and practices are often done in the name of reform but without any data to back them up, we propose that the use of TIMSS data that provides a broad cross-country perspective to influence practice may be the lesser of two evils. The TIMSS data, abide not supportive of causal inferences, does provide interesting comparisons between the United States and other countries. The use of such information for changing or reforming U.S. science education may be more well-grounded in data than what is traditionally done. This is not to advocate such a practice in theory but to suggest that: In the political climate in which we find ourselves today, where immediate action is sought basing policy on such comparative data given how broad the context is (40 other countries) may be a good way to proceed. This is especially true when U.S. practice is demonstrated to be the statistical outlier when contrasted with the other 40 some countries, and given our relatively poor performance on the achievement side.

CURRICULUM COHERENCE

Previous international studies have not always elucidated the nature of the curricula in the countries they examined, leaving much to speculation. In TIMSS, given an analysis of the official documents and textbooks together with information from teachers as to what was actually implemented allowed us to generate a fairly detailed profile as to the nature of the science curriculum in each of these different countries (Schmidt et al., 1999; Schmidt, McKnight, & Raizen, 1997; Schmidt et al., 1997).

When examining these data (together with the achievement data) one of the first conclusions that emerges is that the curriculum is an important predictor of academic learning in eighth grade (Schmidt et al., 1999; Schmidt, McKnight, Houang, Wang, Wiley, Cogan, & Wolfe, 2001). Two features pertaining to the curriculum emerged as being particularly important. The first of these is what we have termed coherence and the second has to do with the rigor of the content. Here we discuss coherence and leave rigor for the following section.

Coherence refers to the degree to which the science curriculum within a country is constructed so as to be conceptually integrated. A sufficiently coherent curriculum provides students with an educational opportunity to understand science in its various constituent parts and, just as importantly, to see how the parts fit together to create a still broader set of notions pertaining to understanding the natural world in which we live. Schmidt and others (1998; 1999) in their studies of the U.S. curriculum found, instead of coherence, disconnected, unfocused, highly repetitive contents resulting in false dichotomies such as the familiar one that pits understanding basic science facts against engaging in empirical work or hands-on science as some would call it. These features of U.S. middle school science are not consistent with the features of science education in most other countries, especially not with those whose students perform best on the achievement test (Valverde & Schmidt, 2000).

Perhaps related to coherence, science education in the United States seems to be built on the principle of "the more topics covered in a given year, the better the resulting learning." This is in marked contrast to that of the top achieving countries where the principle seems to be to focus on a small selective set of topics each year and varying those across the years (Valverde & Schmidt, 2000). The achievement results reflecting learning (the gain from seventh to eighth grade) yield a pattern consistent with the notion that a more focused attention on a fewer number of topics will result in better learning for those topics receiving the focused attention (Schmidt et al., 1999; Schmidt et al., 2001). In fact, the pattern of achievement gains for eighth graders in U.S. science courses reflects little learning taking place in most areas. We found this pattern as we examined the achievement results in 17 areas of science at the eighth grade. The U.S. pattern of gains indicated little learning in any of the 17 areas (Schmidt et al., 1999). This was in marked contrast to patterns of other countries, where large gains were recorded in certain topic areas—those covered during the academic year—with relatively small gains or no gains in areas not covered. Table 2 shows some selective results for some of the 17 areas for the 40 countries involved in the Population 2 part of TIMSS.

Another feature of coherence has to do with the superstructure by which science education is organized. This harks back to the argument on the organization of the course offerings. Chemistry and physics have become an integral foundation for understanding biology, for example. A curriculum needs to be structured such that topics of chemistry necessary to deal with the more advanced biology topics, like energy generation and transfer, should precede the coverage of those biology topics. In examining the science course topic organization of the top achieving countries across the first eight grades, one notes a hierarchical structure in terms of the sequence of topics that is consistent with that principle. This is not the case for the U.S. data (Schmidt & Wang, in press).

The practice in many other countries of not segregating the disciplines according to grade level but considering them simultaneously within the same academic year responds directly to the issue of coherence. Such organization does not come about spontaneously, does not simplify life for the teachers, and is not immediately less expensive. So where does the plan for such an organization originate? Perhaps in reflecting on the way the sciences are intertwined. The organization is likely motivated by a careful consideration of the overall content over the long term. This is the essence of coherence.

Table 2. Percentage Points Gained in Specific Science Areas for Eight Grade (Thirteen-year-olds) in Selected Countries (National Mean Percent Gain with Standards Errors in

Earth Processes

- Czech Republic 10.9 (1.9)
- France 9.5 (1.3)
- Slovak Republic 8.7 (1.3)
- Sweden 8.3 (1.3)
- Norway 8.3 (1.4)
- Denmark 7.0 (1.5)
- Austria 6.7 (1.7)
- Belgium (Fr) 5.9 (1.6)
- Thailand 5.9 (1.2)
- Singapore 5.8 (1.9)
- New Zealand 5.7 (1.5)
- Portugal 5.3 (1.2)
- Australia 5.3 (1.4)
- England 5.2 (1.4)
- Iceland 5.0 (2.0)
- Greece 4.8 (1.0)
- Spain 4.8 (1.1)

Internal Mean 4.8

- Switzerland 4.7 (1.3)
- Hungary 4.7 (1.3)
- Romania 4.6 (1.7)
- Ireland 4.5 (1.5)
- Scotland 4.4 (1.5)
- Lithuania 4.3 (1.5)
- Slovenia 4.2 (1.1)
- Japan 4.0 (0.9)
- Germany 4.0 (1.7)
- Colombia 3.8 (1.7)
- Hong Kong 3.8 (1.7)
- Bulgaria 3.6 (2.0)
- Canada 3.1 (1.3)
- Russia Federation 3.0 (1.3)
- Cyprus 3.0 (1.3)

USA **2.7 (1.7)**

- Korea 2.7 (1.0)
- Latvia 2.6 (1.4)
- Belgium (Fl) 2.6 (2.2)
- Netherlands 2.4 (2.1)
- Philippines 2.2 (1.6)
- Iran 2.2 (1.4)
- South Africa 0.0 (1.8)

Diversity & Structure of Living Things

- Lithuania 11.7 (1.4)
- Japan 9.8 (0.6)
- Russia Federation 9.5 (1.4)
- Portugal 8.8 (1.0)
- Sweden 8.1 (1.1)
- Singapore 7.8 (1.7)
- Latvia 7.6 (1.3)
- Scotland 6.9 (1.4)
- Belgium (Fr) 6.6 (1.3)
- Ireland 6.5 (1.4)
- New Zealand 6.4 (1.3)
- Hungary 6.3 (1.1)
- Cyprus 6.1 (1.0)
- Iceland 6.1 (1.5)
- France 6.1 (1.1)
- Australia 5.9 (1.2)
- Denmark 5.5 (1.2)
- England 5.4 (1.1)
- Norway 5.3 (1.3)

International Mean 5.2

- Spain 5.2 (0.8)
- Austria 5.1 (1.2)
- Germany 5.1 (1.5)
- Greece 5.1 (1.0)
- Switzerland 5.0 (1.1)
- Bulgaria 5.0 (2.0)
- Czech Republic 5.0 (1.2)
- Netherlands 4.5 (1.6)
- Korea 4.4 (1.0)
- Slovak Republic 4.3 (1.0)
- Colombia 3.1 (1.5)
- Canada 3.0 (0.9)
- Slovenia 3.0 (1.1)
- Romania 2.7 (1.4)
- Hong Kong 2.6 (1.5)
- Thailand 2.5 (1.2)
- Iran 2.4 (1.1)

USA **2.0 (1.5)**

- Philippines 1.7 (1.4)
- Belgium (Fl) 1.2 (1.5)
- South Africa 0.3 (1.7)

Life Cycles & Genetics

- Singapore 10.3 (2.1)
- Lithuania 10.1 (2.0)
- France 8.0 (1.4)
- Latvia 7.9 (1.7)
- Belgium (Fr) 7.9 (1.8)
- England 7.5 (2.2)
- Ireland 7.4 (1.8)
- Russia Federation 7.3 (1.5)
- Cyprus 7.1 (1.7)
- Iceland 7.1 (2.1)
- Scotland 7.1 (1.8)
- Hungary 7.0 (1.5)
- Australia 6.9 (1.5)
- Sweden 6.9 (1.3)
- Thailand 6.1 (1.6)
- Slovenia 6.0 (1.6)
- New Zealand 5.9 (1.7)
- Denmark 5.8 (1.6)
- Spain 5.6 (1.4)

USA **4.9 (1.6)**

International Mean 4.8

- Romania 4.8 (2.0)
- Netherlands 4.7 (2.4)
- Portugal 4.7 (1.4)
- Czech Republic 4.3 (1.5)
- Germany 3.7 (1.8)
- Canada 3.6 (1.1)
- Bulgaria 3.4 (2.8)
- Greece 3.3 (1.5)
- Norway 3.3 (1.5)
- Korea 3.0 (1.3)
- Switzerland 2.9 (1.4)
- Slovak Republic 2.8 (1.2)
- Hong Kong 2.8 (1.9)
- Austria 2.1 (1.7)
- Colombia 1.9 (2.1)
- Iran 1.4 (2.7)
- Belgium (Fl) 0.4 (1.8)
- Japan 0.1 (1.0)
- Philippines −0.4 (1.4)
- South Africa −2.3 (1.8)

Forces & Motion

- Lithuania 13.6 (2.2)
- Netherlands 11.9 (2.0)
- Greece 10.0 (1.6)
- Sweden 9.4 (1.6)
- Latvia 9.2 (2.0)
- Scotland 8.8 (2.2)
- Portugal 8.6 (1.6)
- Iran 8.5 (2.2)
- Spain 8.3 (1.5)
- Cyprus 8.0 (1.6)
- Switzerland 7.1 (1.6)
- Colombia 6.8 (2.2)
- France 6.5 (1.8)
- Singapore 6.5 (1.5)
- New Zealand 6.4 (1.6)
- England 6.0 (2.0)
- Australia 5.9 (1.6)
- Germany 5.9 (2.1)
- Slovenia 5.8 (1.7)

International Mean 5.6

- Ireland 5.6 (1.9)
- Norway 5.5 (2.1)
- Hungary 5.5 (1.8)
- Austria 5.4 (1.8)
- Philippines 5.3 (1.6)

USA **5.0 (2.1)**

- Korea 4.8 (1.4)
- Hong Kong 4.5 (2.1)
- Romania 3.9 (1.9)
- Canada 3.8 (1.7)
- Belgium (Fr) 3.6 (2.1)
- Japan 3.3 (1.2)
- Bulgaria 3.1 (2.9)
- Denmark 2.8 (1.9)
- Thailand 2.7 (1.7)
- Slovak Republic 2.5 (1.6)
- South Africa 2.1 (2.0)
- Czech Republic 2.0 (1.5)
- Russia Federation 1.7 (1.8)
- Iceland 0.9 (3.0)
- Belgium (Fl) −2.7 (3.0)

Structure of Matter

Country	Score (SE)
Lithuania	25.0 (1.8)
Singapore	24.2 (2.0)
Russia Federation	19.5 (2.4)
Greece	18.4 (1.3)
Sweden	18.2 (1.5)
Portugal	17.2 (1.4)
Latvia	15.3 (1.8)
Bulgaria	13.3 (3.0)
Spain	12.5 (1.6)
Australia	11.6 (1.5)
Norway	11.1 (1.4)
Thailand	10.5 (1.5)
Iran	10.5 (1.6)
Ireland	10.1 (1.8)
Netherlands	9.6 (1.9)
Internal Mean	9.1
New Zealand	9.0 (1.6)
Germany	8.9 (1.9)
Japan	8.6 (1.3)
England	7.9 (1.9)
USA	**7.9 (1.6)**
Romania	7.9 (2.0)
Slovak Republic	7.5 (1.8)
Switzerland	7.3 (1.2)
France	6.7 (1.5)
Korea	6.6 (1.3)
Czech Republic	6.3 (2.4)
Cyprus	6.1 (1.3)
Denmark	5.9 (1.5)
Austria	5.5 (2.0)
Scotland	4.6 (1.8)
Canada	4.2 (1.3)
Philippines	4.1 (1.3)
Hungary	3.9 (1.7)
Iceland	3.8 (1.6)
Belgium (Fr)	3.6 (1.6)
Colombia	2.6 (2.2)
Hong Kong	1.5 (1.6)
South Africa	0.9 (1.2)
Belgium (Fl)	0.4 (1.4)
Slovenia	-0.4 (1.7)

Energy & Physical Processes

Country	Score (SE)
Russia Federation	8.9 (1.2)
Latvia	8.6 (0.9)
Lithuania	8.4 (0.9)
Portugal	8.2 (0.7)
Greece	8.0 (0.7)
Netherlands	7.2 (1.4)
Cyprus	7.0 (0.6)
Singapore	6.8 (1.3)
Czech Republic	6.8 (1.1)
Denmark	6.7 (1.0)
Norway	6.6 (0.8)
Iran	6.6 (1.0)
New Zealand	6.6 (1.0)
Scotland	6.5 (1.1)
Slovak Republic	6.3 (0.9)
Austria	6.1 (1.1)
Ireland	6.1 (1.1)
Switzerland	5.9 (0.7)
France	5.8 (0.8)
Slovenia	5.8 (0.8)
Belgium (Fr)	5.5 (0.9)
Internal Mean	5.4
Hungary	5.3 (0.8)
Australia	5.3 (1.0)
Spain	5.1 (0.7)
Sweden	4.9 (0.8)
Romania	4.8 (1.1)
Canada	4.6 (0.7)
Iceland	4.3 (1.2)
England	4.3 (1.0)
Bulgaria	4.3 (1.5)
Japan	3.7 (0.5)
Hong Kong	3.6 (1.5)
Thailand	3.6 (1.0)
USA	**3.6 (1.3)**
Germany	3.4 (1.2)
Belgium (Fl)	3.2 (1.2)
Colombia	2.4 (1.2)
Korea	2.2 (0.7)
Philippines	1.8 (1.2)
South Africa	1.7 (1.7)

Physical Changes

Country	Score (SE)
France	14.8 (2.4)
Lithuania	14.3 (2.7)
Portugal	13.6 (1.8)
Spain	12.5 (2.0)
New Zealand	11.3 (2.1)
Greece	10.3 (1.6)
Russia Federation	10.2 (2.7)
Norway	9.9 (2.7)
Hungary	9.7 (2.0)
Iceland	9.3 (3.4)
Iran	9.0 (2.5)
Sweden	8.9 (1.9)
Latvia	8.9 (2.3)
Austria	8.6 (2.4)
Denmark	8.4 (2.4)
Cyprus	8.3 (2.4)
Netherlands	7.9 (2.8)
Slovak Republic	7.8 (2.0)
Slovenia	6.8 (2.2)
Internal Mean	6.2
USA	**5.9 (2.4)**
Ireland	5.2 (2.4)
Czech Republic	5.1 (2.5)
Canada	5.1 (2.3)
Germany	4.7 (2.5)
Philippines	4.5 (2.0)
Thailand	4.5 (1.9)
Belgium (Fr)	4.3 (2.4)
Romania	4.1 (2.0)
Colombia	3.6 (2.3)
Switzerland	3.2 (1.8)
Australia	2.5 (1.9)
Japan	2.0 (1.7)
Belgium (Fl)	1.6 (2.4)
South Africa	1.6 (2.4)
Scotland	1.3 (2.2)
England	1.1 (2.7)
Hong Kong	0.7 (2.3)
Bulgaria	0.1 (4.8)
Singapore	-0.4 (2.1)
Korea	-3.1 (1.8)

Chemical Changes

Country	Score (SE)
Lithuania	19.0 (1.5)
Russia Federation	17.2 (2.0)
Portugal	16.4 (1.3)
Japan	13.4 (0.7)
Latvia	13.1 (1.4)
Scotland	11.4 (1.7)
New Zealand	10.6 (1.3)
Korea	10.4 (1.1)
Slovak Republic	9.7 (1.3)
Cyprus	9.3 (1.1)
Spain	9.2 (1.1)
Sweden	8.8 (1.0)
Austria	8.8 (1.5)
Czech Republic	8.6 (1.5)
Greece	8.5 (1.0)
Netherlands	8.5 (1.3)
Belgium (Fl)	8.1 (1.3)
Internal Mean	7.9
Hungary	7.8 (1.1)
Australia	7.7 (1.2)
Germany	7.7 (1.6)
Switzerland	7.4 (1.1)
England	7.1 (1.5)
Iran	6.7 (1.4)
France	6.5 (1.4)
Ireland	6.4 (1.5)
Canada	6.1 (1.1)
Norway	5.9 (1.2)
Iceland	5.5 (1.4)
Thailand	5.4 (1.5)
Denmark	5.4 (1.1)
Singapore	5.4 (1.8)
Bulgaria	5.2 (1.9)
Hong Kong	5.2 (1.8)
USA	**5.0 (1.6)**
Romania	4.3 (1.5)
Philippines	4.1 (1.4)
South Africa	3.9 (1.8)
Slovenia	3.8 (1.4)
Belgium (Fr)	3.0 (1.2)
Colombia	0.6 (1.6)

In contrast the U.S. practice of segregating the disciplines by grade
level and then introducing topics into the curriculum without regard for the
hierarchical structure that is necessary for moving to deeper levels of
knowledge does not indicate the use of coherence as an organizing principle for curriculum. Perhaps this is reflective of the fact that most middle
school science in this country is descriptive in nature, focusing on various
features of the earth and solar system as well as on descriptive features of
the plant, animal, and, especially, human worlds. For the middle school curriculum in other countries, the coverage of topics in earth science and
biology is deeper, in part because it is predicated on knowledge of chemistry and atomic structure (Valverde & Schmidt, 2000; Schmidt, *et al.*, 2001).

Table 3 shows data that speak to this issue. The table indicates which
topics are introduced at which grade level in the United States as contrasted
with the top achieving countries (A+ countries). The A+ countries[2] represent that select set of four countries whose children performed at the
highest levels for the eighth grade on the overall science test. Three grade
spans are used for this table. The obvious conclusion from this table is that
the nature of the U.S. curriculum is organized quite differently from those
of the top achieving countries.

There is a somewhat hierarchical structure undergirding the introduction of topics in the A+ countries that is obviously absent from the
United States curriculum. In the United States, the principle seems to be to
introduce almost everything at an early point—while in other countries concepts are introduced sequentially—some only after other prerequisites have
been introduced in order to instill a deeper understanding of the subsequent topics.

A good example to illustrate this point is the topic of cells. In the
United States this topic is introduced in the elementary school, but in
the A+ countries this topic is first introduced only at the fourth to
sixth grade levels. This is probably because the topics of atomic structure
and chemical properties are necessary at one level to understand in a
deep way what the cell is, how it functions, and how it carries on its life
processes. Introducing cells in the elementary school clearly is done on a
different level and may seem innocuous on the surface. Yet again one has
to question the U.S. approach compared to that of other countries,
especially when our levels of achievement and learning associated with
science are not considered acceptable. The data confirming such unaccept-
ability does not just come from international studies like TIMSS but also

---

[2] Singapore, Czech Republic, Japan, and Korea are the A+ countries in the TIMSS science
study.

Table 3. Science Topics Intended for Introduction in Various Stages[a]: TIMSS Top Achieving Countries (A+) and the United States

Topics Intended for Introduction in Grades 1 to 3		Topics Intended for Introduction in Grades 4 to 6		Topics Intended for Introduction in Grades 7 to 8	
by at least 50% of the A+ Countries	by at least 50% of the States	by at least 50% of the A+ Countries	by at least 50% of the States	by at least 50% of the A+ Countries	by at least 50% of the States
Earth's composition				Earth's composition	Biochemical processes in cells
Land forms				Land forms	
Bodies of water		Bodies of water		Physical cycles	
Atmosphere		Atmosphere		Microorganisms	
Rocks, soil		Rocks, soil		Organism energy handling	
Weather and climate		Weather and climate	Building and breaking	Reproduction of organisms	
Physical cycles			Earth's history	Human nutrition	
Earth in the solar system		Earth in the solar system			
		Planets in the solar system		Atoms, ions, molecules	
Planets in the solar system		Beyond the solar system	Beyond the solar system	Subatomic particles	
Plants, fungi		Plant, fungi		Energy types, sources, conversions	
Animals		Animals		Explain physical changes of matter	
Microorganisms		Organs, tissues		Math, science influence on society	
Organs, tissues		Cells		Pollution	
Cells		Organism sensing, responding	Reproduction of organisms	Material and energy res. conservation	
Organism energy handling		Life cycles of organisms			
Organism sensing, responding		Biomes, ecosystems			
Life cycles of organisms		Habitats and niches			
Variation and inheritance		Interdependence of living things			
Evolution, speciation, diversity					
Biomes and ecosystems			Human disease		
Habitats and niches			Biochemistry of genetics		
Interdependence of living things			Atoms, ions, molecules		

Table 3. *Continued*

Topics Intended for Introduction in Grades 1 to 3		Topics Intended for Introduction in Grades 4 to 6		Topics Intended for Introduction in Grades 7 to 8	
by at least 50% of the A+ Countries	by at least 50% of the States	by at least 50% of the A+ Countries	by at least 50% of the States	by at least 50% of the A+ Countries	by at least 50% of the States
Human nutrition		Classification of matter			
Classification of matter		Physical properties of matter			
Physical properties of matter		Chemical properties of matter			
Chemical properties of matter		Heat and temperature			
Energy types, sources, conversions		Sound and vibration			
Heat and temperature		Light	Wave phenomena		
Sound and vibration					
Light		Electricity			
Electricity		Magnetism	Magnetism		
Types of forces		Physical changes	Physical changes		
Time, space, and motion		Chemical changes	Chemical Changes		
Nature or conceptions of technology		Types of forces			
Interactions of sci., tech. and society		Time, space, motion	Dynamics of motion		
History of science and technology		Sci. applications in math, tech.	Interactions of sci, math, and tech.		
Land, water, sea res. conservation					
Material and energy res. conservation		Land, water, sea res. conservation	Pollution		
Nature of scientific knowledge		Science and mathematics	World population		
			Food production, storage		
			The scientific enterprise		

[a] Certain science framework topics were typically introduced at various stages of schooling, grades 1–3, 4–6 and 7–8. This table presents a broad picture of the typical sequence of science topic introductions (representing the aggregate) in the A+ countries: Singapore, Czech Republic, Japan, and Korea, and in the 21 states.

from the National Assessment of Educational Progress (NAEP) and other state assessments.

The last issue to be raised under the banner of coherence has to do with the false dichotomies that seem to propagate in the science education field. As mentioned in the section reviewing the history of American science education reform, the definition of science instruction has always been a contentious issue. The dichotomy that suggests there is basic factual science on the one hand and hands-on or inquiry science on the other, is, we argue, a false one and reflects education jargon created around ideology rather than recognizing the inherent coherence of the science disciplines. In many other nations, data would seem to imply that both content and inquiry are integral parts of science education. There is no dichotomy giving one precedence over the other. Rather, it would seem that when focused instruction is provided, that instruction entails not only covering the basic scientific material and understandings necessary to conduct empirical inquiry but also involves empirical inquiry so that students have first-hand experience with data.

Doing the latter without the former seems mere busywork for students (content-neutered process), but doing the former without the latter also seems sterile and without context. In short, the question is not "which one" but "what is the balance" between the two for providing a coherent picture of science.

WORLD-CLASS STANDARDS

The criterion of coherence becomes more important as course content becomes more complex and the expectations for learning are more demanding. This brings us to the final focus of the use of TIMSS for an evaluation, the actual substance of science instruction in the middle grades compared to that of other countries—what we termed as the issue of rigor. As observed in Table 3, U.S. middle school science instruction seems to focus on the descriptive aspects of the earth and biological sciences, with some concern for environmental issues. By contrast, the curricula of other countries during the middle grades focus on the fundamentals of chemistry and physics and the more advanced topics in biology and earth science that are based on and derived from the principles in chemistry and physics (Schmidt *et al.*, 1999; Valverde & Schmidt, 2000; Schmidt & Wang, in press). As such, the level of the topics, from a scientific disciplinary point of view, is much more advanced in other countries than it is in the United States. This seems to be especially important for U.S. science educators to investigate. The absence of most of the serious physics and chemistry topics for the seventh and eighth grade in the United States stands so starkly in contrast with science courses for the same grades in the top achieving countries—in fact with those of most other TIMSS countries.

One of the observations made with respect to TIMSS achievement data was that the longer one goes to school in America, the lower one sinks in international ranking in a relative sense. The initial decline in science achievement from fourth grade to eighth grade (now confirmed by TIMSS-R for the same cohort of students) is particularly acute as the U.S. ranking went from being tied for second in the world, outperformed statistically only by South Korea, to essentially being middling in performance and only slightly above the international mean by eighth grade. That precipitous decline in relative standing is more interpretable and understandable when one looks at the nature of science instruction for the middle grades in the United States and then investigates how science is taught in other countries.

These data are particularly alarming when one combines the information from this study, which shows that most U.S. students do not get serious instruction in physics and chemistry topics in the middle school, with the fact that only about 25 percent of U.S. high school students ever take a physics class (National Center for Education Statistics, 1998). The implication is that most U.S. high school graduates have had no serious introduction to the natural laws of the physical world in which they live. Yet researchers of scientific literacy describe how important it is for the American public to be able to respond to issues pertaining to the allocation of resources. From simple issues dealing with energy to more complex issues such as the scientific notions behind global warming, from current research on cyclotrons and the fundamental particles of physics to inorganic chemical reactions, from nuclear power issues to the nature of fundamental scientific research—all of these topics would require relatively sophisticated understanding of some physics and chemistry, yet the United States seems not to provide most students with an opportunity to acquire such understandings.

THE METHODOLOGY OF TIMSS APPLIED TO OTHER EVALUATIONS

We have argued in the previous sections that TIMSS was never designed to provide evaluation data with respect to particular science programs or with respect to particular reforms. We have also argued that TIMSS can be used in an evaluative capacity to examine certain broad features of science education especially curriculum design and to suggest hypotheses for reform.

The question we address in this section focuses on how the conceptual design and basic methodology of TIMSS can be used in more specific science program evaluations. In the following section, we draw on TIMSS methodology to collect other data with respect to issues of science curriculum evaluation. New data collection using TIMSS methodology will again illuminate our arguments for how to assess broad features of science education.

The Measurement of Achievement

In order to use TIMSS methodology for science education evaluation some general principles governing collection of the achievement data need to be stated. From the TIMSS study we learned that total scores summed over items from multiple topics within science are not particularly curriculum sensitive (Schmidt, Jakwerth, & McKnight, 1998). Allowing countries to pick items that would best represent their curriculum but picked across the four areas of science (biology, chemistry, geology, and physics) produced an array of scores by which the countries could be ranked. The amazing result was that, on any of these overall total scores, countries essentially ranked the same and did not change in their relative standing except in minor ways.

When more specific areas of the curriculum were used to formulate subtests, country rankings altered appreciably, reflecting differences in curricula. We call such differences "curriculum" sensitive measurement. The first general principle suggested for science education evaluation in the United States is that tests used to evaluate science learning should always be focused on very specific topic areas. Multiple test scores should be used as a profile to characterize the learning that has taken place, not total science scores.

The second general principle is to design a "pseudo longitudinal" study so as to measure learning instead of achievement status. Although longitudinal studies are preferred, they often are too costly and difficult to realize in which case a pseudo-longitudinal study can be done by measuring two adjacent age groups. In TIMSS we created such a quasi-longitudinal study at least at the country level be measuring the two adjacent grades associated with a particular age cohort (e.g., 13 years old leading to grades seven and eight in the U.S.). By using the lower of the two grades as a surrogate pre-measure for the upper grade, we could examine the effects of science instruction in the eighth grade, having accounted for the learning that had taken place up through the seventh grade. This second principle allows for an understanding relative to a single grade rather than trying to characterize the educational experience of the student up to and including that grade level in question. This approach of isolating the data that reflect learning in a specific year rather than the cumulative learning to date is critical in an evaluation study.

Using the TIMSS Conceptual Framework for Evaluating Science Education Programs

As described earlier, the TIMSS study was based on a conception of education systems as providing potential opportunities for students to learn. Every aspect of the data collection, including the questionnaires and the curriculum analysis, was designed with this as the central focus. Thus the TIMSS study rests on a conceptual framework that treats the education system as a whole. Development of the TIMSS test was also based on this conception and allowed for, as mentioned previously, the measurement of achievement gains in specific areas of the curriculum for which other data were collected such as from teachers and curricular materials. Figure 1 shows this conceptual model with four fundamental questions guiding the inquiry. The questions focus on what an education system is, starting with what students are expected to learn, who delivers the associated instruction, and how that instruction is finally delivered.

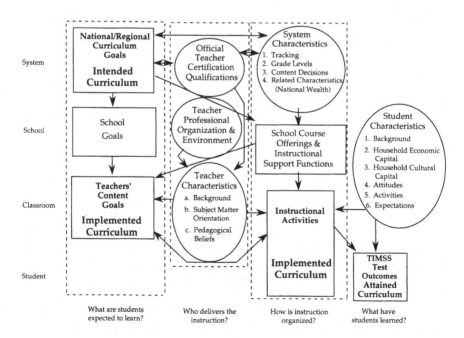

Figure 1. The TIMSS Conceptual Framework.

We suggest using this model as a basis for doing curriculum evaluation including modifying TIMSS questionnaires or developing new ones based on the model we are currently using this general model to evaluate science education in two contexts. We briefly describe each of those, not to represent the outcomes or the results since both studies are only recently under way but rather to suggest concretely two examples of how the general system's model and associated instruments can be used in the conduct of science program evaluation.

CASE ONE: TIMSS-REPEAT BENCHMARKING STUDY

TIMSS-Repeat (TIMSS-R) was designed as a follow-up study for the original TIMSS study (Martin *et al.*, 2000; Mullis *et al.*, 2000). The data, collected in 1999 on eighth graders, represented the same cohort of students who in TIMSS had been fourth graders. The basic idea behind the TIMSS-R was to see whether changes took place over the four years for those fourth grade students. In one sense, TIMSS-R is a cohort longitudinal study.

As a part of the larger TIMSS-R study (the TIMSS-R Benchmarking Study), several districts, district consortia, and states participated by drawing large and representative samples in a way so as to be considered as a quasi country in the cross-national comparisons of TIMSS-R. This is similar to what the First in the World Consortium and Minnesota did in the original TIMSS study. Around 25 of these groups participated fully in the TIMSS-R data collection, including questionnaire administration and achievement testing. Through the support of the National Science Foundation (NSF), a small number of those districts were selected for further study to examine their systemic efforts and its impact on achievement. The selected districts included systemic initiative sites and curriculum development sites funded by the NSF.

Here we will refer not to that entire benchmarking study but only to the work done with the select set of six groups. For these six groups additional questionnaires were developed, based on the model of potential educational opportunities represented in Figure 1. The teachers who had participated in the original TIMSS-R study from these districts were requested to respond to the Teacher Questionnaire.

A District Questionnaire was also administered addressing issues that had been discovered in TIMSS to play a major role in the model outlined in Figure 1. This additional data collection enabled us to make the study more like a curriculum evaluation by focusing on specific district programs. That is, a district or a consortium of districts enables us to focus the evaluation on the particular science education program as defined by that district or consortium. This in one sense comes closer to the type of science curriculum evaluation study that is more traditionally done. Since the program of study in a given district is clearly articulated, using the system model enables us to collect information pertaining to the structures that are in place to support that particular implementation. With the TIMSS-R achievement results it is possible to associate the profile of learning that takes place in that district with the features of the science curriculum in that district or consortium.

The only difficulty in all of this is that the TIMSS-R study does not provide the kind of quasi longitudinal data called for by the second principle mentioned in the methodology section. Therefore, it will be more difficult to make analyses comparable to those made in TIMSS, because one will have to examine the curriculum and the educational experiences of the students through their first eight years of school rather than focusing simply on the eighth grade. Given the narrow scope of the entity being studied (a district as opposed to a country as a whole) and the additional data collection, we anticipate being able to relate various features of the science program to student performance on the achievement tests. This study is currently underway, and the results will be available in 2001.

CASE TWO: HANDS-ON-UNIVERSE STUDY

This example actually exemplifies a typical program evaluation in science education. The Hands-on Universe (HOU) is an empirically driven program designed at the University of California, Berkeley; using data downloaded by the Hubbell telescope and various other telescopes around the world, HOU is intended for use in teaching physics and astronomy (see http://hou.lbl.gov). A salient feature of this program, as with many current science programs, is to engage students in actual empirical inquiry. The images downloaded from various telescopes are provided through a computer program with software that enables students to explore the universe in conjunction with studying phenomena such as black holes and red shifts as well as identifying celestial objects not yet identified.

This program is designed to bridge the gap in the artificial dichotomy discussed earlier between the "substance of science" and "hands-on science." The attempt here is to provide a good, sound, scientific basis on several astrophysics principles as well as on the measurement of the universe from an astronomical point of view. A parallel focus is to provide an engaging task where students are actually involved in the analysis of real data. As such, the evaluation question is, does this program achieve its stated objectives? In other words, do students who are taught physics or astronomy through the HOU materials learn the scientific principles involved as well as to acquire inquiry skills in physics and astronomy?

Our involvement with HOU was to design and provide an evaluation by which to answer the above questions. This particular task is quite different from both the original TIMSS study and the TIMSS-R Benchmarking study described in the previous section. In both of those cases the focus of the evaluation is the system of education and how that system provides science education. In this particular application, the evaluation involves examining the particulars of a specific program applied in multiple school contexts and in different education systems.

The model upon which this evaluation was designed is the same model referenced in Figure 1. Using the same general approach with regard to the nature of the provision of potential educational experiences, many of the TIMSS questionnaires could be appropriately used within this same context. This was true especially of the school, teacher, and student questionnaires. None of the achievement tests, however, would serve the purpose here since the TIMSS and the TIMSS-R tests were designed as general science tests instead of tests of astrophysics.

In order to accommodate those differences, the first task was to revise the TIMSS science framework, which was the underpinning of most of the previous work done in TIMSS. The resulting HOU conceptual framework specified 87 distinct topics from chemistry, physics, and earth science. This new framework was similar to the TIMSS curriculum framework in many aspects (Schmidt *et al.*, 1997), which involved not only content topics but also the performance expectations associated with the topics.

This particular modification of the conceptual framework was done because a large portion of the original TIMSS curriculum framework was mostly irrelevant to the HOU program. The TIMSS framework included a specification of topics associated with biology and environmental sciences. None of these would be germane to HOU, although some of the earth science, chemistry, and physics topics were. Thus, the particular focus of this evaluation on astrophysics and astronomy demanded a further delineation of those content topics within the TIMSS curriculum framework. To accomplish this, scientists working on the project were engaged to further expand the TIMSS framework for the HOU project. Additional topics were added to the framework to delineate astronomy and astrophysics. This was done such that those topics were nested within the current framework, which would thus allow the collapsing of those categories to the same topics in the original TIMSS framework. This allows the data to be comparable with data from TIMSS and TIMSS-R.

The resulting revised science curriculum framework was then used as the basis for revising the questionnaires and developing an HOU achievement test. For example, the basic TIMSS teacher questionnaire was used but the particular items asking teachers about what they taught were aimed at the newly revised topics in astrophysics and astronomy.

An HOU achievement test had to be developed completely from scratch using standard test development procedures. In the process of developing the HOU achievement tests, we did incorporate several TIMSS mathematics and physics items, which were conceived as relevant to the study of astrophysics and astronomy. The rationale is that these items would allow an anchoring against the original TIMSS test.

The use of TIMSS as a conceptual framework to design an evaluation of HOU suggests the applicability of the methods and the conceptual framework to other kinds of program evaluations in science education. Being system focused, the framework and procedures provide a broad sweep of data that is germane to the evaluation of any particular science program. Conversely, the particulars of any specific science program would need to be incorporated as was done for HOU. This evaluation of the HOU project has begun, with the initial data being collected during the 2000–2001 academic year.

FINAL THOUGHTS

The main question we wished to address in this paper is what role TIMSS should play in the evaluation of science education in the United States. We think it should play a limited role. In further delineating the answer to that question, a key distinction is between TIMSS as a study and TIMSS as a set of procedures and methodologies. We argue that TIMSS was not designed to provide a specific evaluation about science reform in the United States or about particular programs. Thus, TIMSS should not be used in that fashion.

On the other hand, we have argued that TIMSS could be used as a mechanism to generate hypotheses and perhaps even policies, about the current general structure and form of science education in the United States as compared with other countries. We argue that such issues as the organizational structure of science education offerings, the degree of coherence integral to those offerings, and the rigor of the actual science being taught are all areas where data from TIMSS could be used cogently to make hypotheses on how science education might be reformed. At the very least they would provide hypotheses that could be examined and further researched. We further argue that, given the current dissatisfaction with science education, these empirical results would certainly provide a limited basis for making policy and practice changes, especially in light of the fact that many policies and practices reflect positions backed up by ideology much more than positions backed up by data.

Furthermore, we propose that the conceptual framework and its associated methodologies can serve as the basis for conducting meaningful evaluation in science education. Two cases in which we are currently engaged are provided as examples of such applications. The key feature is the evaluation model which is based on the notion that educational programs are designed to provide students with opportunities to learn and that the quality of those opportunities affects what students subsequently will learn. The model and its associated instruments were designed to measure those opportunities and to relate them to achievement. These principles, as stated, surely serve as a foundation for program evaluation in science education.

The final proof concerning the use of TIMSS conceptions and methodologies for doing science program evaluation awaits the results of the two studies—TIMSS-R Benchmarking and the HOU project. When results become available, the field may judge whether program evaluation based on the TIMSS model is a productive way to think about the problems, and correspondingly, to collect data relative to those issues. We await those results and its judgement.

REFERENCES

American Association for the Advancement of Science. (1993). *Benchmarks for Science Literacy*. Oxford University Press, New York.

DeBoer, G. E. (1991). *A History of Ideas in Science Education: Implications for Practice*. Teachers College Press, New York.

Kyle, W. C. (1991). Curriculum development projects of the 1960's. *Research within Research: Science Education*, pp. 3–24.

Martin, M. O., Mullis, I. V. S., Gonzalez, E. J., Gregory, K. D., Smith, T. A., Chrostowski, S. J., Garden, R. A., and O'Connor, K. M. (2000). *TIMSS 1999 International Science Report: Findings from IEA's Repeat of the Third International Mathematics and Science Study at the Eighth Grade*. International Study Center, Boston College, Boston, MA.

Mascolo, R. (1969). Performance in conceptualizing: relationship between conceptual framework and skills of inquiry. *Journal of Research in Science Teaching* 6:29–35.

Mullis, I. V. S., Martin, M. O., Gonzalez, E. J., Gregory, K. D., Garden, R. A., O'Connor, K. M., Chrostowski, S. J., and Smith, T. A. (2000). *TIMSS 1999 International Mathematics Report: Findings from IEA's Repeat of the Third International Mathematics and Science Study at the Eighth Grade*. International Study Center, Boston College, Boston, MA.

National Commission on Excellence in Education. (1983). *A Nation at Risk: The Imperative for Educational Reform*. Government Printing Office, Washington, DC.

National Center for Education Statistics. (1998). *Pursuing Excellence: A Study of US Twelfth-Grade Mathematics and Science Achievement in International Context*. NCES 98-049. Government Printing Office, Washington DC.

National Research Council. (1996). *National Science Education Standards*. National Academy Press, Washington DC.

National Science Teachers Association. (1982). *Science-Technology-Society: Science Education for the 1980's*. NSTA, Washington DC.

Quattropani, D. J. (1977). *An Evaluation of the Effect of Harvard Project Physics on Student Understanding of the Relationships Among Science, Technology, and Society*. Ph.D. dissertation, University of Connecticut.

Schmidt, W. H., Jakwerth, P. M., and McKnight, C. C. (1998). Curriculum-sensitive assessment: Content *does* make a difference. *International Journal of Educational Research* 29:503–527.

Schmidt, W. H., and McKnight, C. C. (1995). Surveying educational opportunity in mathematics and science: An international perspective. *Educational Evaluation and Policy Analysis* 17(3):337–353.

Schmidt, W. H., McKnight, C., Cogan, L. S., Jakwerth, P. M., and Houang, R. T. (1999). *Facing the Consequences: Using TIMSS for a Closer Look at US Mathematics and Science Education*. Kluwer Academic Publisher, Dordrecht/Boston/London.

Schmidt, W. H., McKnight, C. C., Houang, R., Wang, H. A., Wiley, D., Cogan, L., and Wolfe, R. (2001). *Why Schools Matter: Using TIMSS to Investigate Curriculum and Learning*. Jossey-Bass, New York.

Schmidt, W. H., McKnight, C. C., and Raizen, S. A. (1997). *A Splintered Vision: An Investigation of U.S. Science and Mathematics Education*. Kluwer Academic Publishers, Dordrecht/Boston/London.

Schmidt, W. H., Raizen, S. A., Britton, E. D., Bianchi, L. J., and Wolfe, R. G. (1997). *Many Visions, Many Aims, Volume 2: A Cross-national Investigation of Curricular Intentions in School Science*. Kluwer Academic Publisher, Dordrecht/Boston/London.

Schmidt, W. H., and Wang, H. A. (in press). Pursuing excellence for the U.S. mathematics and science education standards. Paper Accepted for publication in the *Journal of Curriculum Study*.

Valverde, G. A., and Schmidt, W. H. (2000). Greater expectations: Learning from other nations in the quest for 'world-class standards' in US school mathematics and science. *Journal of Curriculum Study* 32(5):651–687.

Wang, H. A. (1998). *A Content Analysis of the History of Science in the Secondary Science Textbooks*. Ph.D. Dissertation. University of Southern California, Los Angeles.

Welch, W. W. (1973). Review of the research and evaluation program of Harvard Project Physics. *Journal of Research in Science Teaching* 10:365–378.

CHAPTER 3

Evaluating Systemic Reform
Evaluation Needs, Practices, and Challenges

Bernice Anderson

INTRODUCTION

The Directorate for Education and Human Resources (EHR) at the National Science Foundation (NSF) initiated a strategy to create positive changes in K-12 science and mathematics education by means of comprehensive, system-wide efforts coordinated across educational settings. The aim was to engage entire systems (states, cities and districts) to participate in comprehensive endeavors that would elevate teaching and learning standards and enrich the instructional materials and pedagogy offered to students. Central to this effort was the concept that all children can learn high quality mathematics and science.

In 1991 the Statewide Systemic Initiatives (SSI) program was launched. In 1994, NSF expanded its systemic efforts to urban and rural areas through the Urban Systemic Initiatives (USI) and Rural

Bernice Anderson, Division of Research, Evaluation and Communication, National Science Foundation, 4201 Wilson Boulevard, Arlington, VA 22230.

Evaluation of Science and Technology Education at the Dawn of a New Millennium, edited by James W. Altschuld and David D. Kumar, Kluwer Academic / Plenum Publishers, New York, 2002.

Systemic Initiatives (RSI) programs, respectively. Later, the urban systemic reform effort included the Comprehensive Partnerships for Mathematics and Science Achievement (CPMSA) Program for mid-size school districts with large minority student populations, and the current initiative known as the Urban Systemic Program (USP) that gives greater attention to higher education partnerships, the technical and instructional workforce, and the use of technology in support of mathematics and science education.

In general, to be considered *systemic*, an initiative had to successfully describe how they would develop, manage, and evaluate a comprehensive mathematics and science reform effort and operationalize their approach to change by a solid commitment of resources and partnerships to promote and sustain the effort. More specifically, these initiatives required awardees to adopt standards-based curricula and instructional practices; align funding allocations to support high quality science and math instruction; devise and/or implement educational policies so that students have access and motivation to enroll in higher-level courses; and most importantly, document and assess the impact of their efforts on student achievement gains, particularly among disadvantaged student populations. The evaluation findings of the SSI Program indicated that the particular mix of systemic strategies varied due to the needs and challenges facing the state or districts, ideological viewpoints, political or historical traditions, the current policy environment, the locus of influence of the initiative leadership, available resources, and the status and strength of other K-12 reform efforts in facilitating or inhibiting change (Zucker *et al.*, 1998).

Each initiative was required to have its own evaluation. The initiatives also engaged in a mix of program evaluation efforts coordinated by the Foundation and designed for formative and summative purposes. In this chapter an overview of NSF's oversight and accountability efforts will be provided. The major focus will be on third party evaluation efforts, describing the needs being addressed, the evaluation challenges, and future directions.

OVERVIEW OF MONITORING AND ACCOUNTABILITY PRACTICES

Over time, EHR developed an accountability continuum that includes: program/initiative (project) based monitoring, midpoint reviews, program effectiveness reviews, external program evaluations, evaluative research studies, and impact studies. The first three categories may be characterized as internal assessment activities to monitor progress and the latter three as external accountability studies to evaluate program improvements.

Internal Assessment Activities

These oversight strategies were structured to meet annual reporting requirements regarding changes in the systems toward improved outcomes in student learning of mathematics and science, and to inform budgetary decisions. The internal assessment activities were put in place to assess the progress of both the individual initiatives and the systemic initiative programs. To guide the review and assessment of progress, the Division of Educational System Reform (ESR) codified the following six drivers of reform.

Process Drivers

1. The Classroom Driver: Implementation of comprehensive, standards-based curricula as represented in instructional practice, including student assessment, in every classroom, laboratory, and other learning experience provided through the system and its partners
2. The Policy Driver: Development of a coherent, consistent set of policies that supports provision of high quality mathematics and science education for each student; excellent preparation, continuing education, and support for each mathematics and science teacher (including all elementary teachers); and administrative support for all students served by the system
3. The Resource Driver: Convergence of all resources that are designed to support science and mathematics education—fiscal, intellectual, material, curricular, and extra-curricular—into a focused and unitary program to constantly upgrade, renew, and improve the educational program (in mathematics and science) for all students
4. The Stakeholder/Community Driver: Broad-based support from parents, policymakers, institutions of higher education, business and industry, foundations, and other segments of the community for the goals and collective value of the program, based on rich presentations of the ideas behind the program, the evidence gathered about its successes and its failures, and critical discussions of its efforts

Outcome Drivers

5. The Attainment Driver: Accumulation of a broad and deep array of evidence that the program is enhancing student achievement, through a set of indices that might include achievement test scores, higher level courses passed, college

admission rates, college majors, Advanced Placement Tests taken, portfolio assessment, and ratings from summer employers that demonstrate that students are generally achieving at a significantly higher level in science and mathematics

6. The Equity Driver: Improvement in the achievement of all students, including those historically underserved

Monitoring Progress. Evidence of progress is required and monitored yearly. Each site is visited annually by a site visit team led by a program officer from ESR. The purpose is to review the implementation status of the initiative and to examine the planned actions/activities in response to concerns and issues identified by reviewers of the plan being implemented or the experts on site visit teams. In the early phases of the SSI program, site visits by NSF program staff were augmented by site visits conducted by a third party that were designed to provide advice to the Foundation as the systemic reform concept was evolving and being shaped by the local context. Over time, third party site-based monitoring efforts transitioned to a distance monitoring system. More specifically, an annual web-based collection of core data elements has been used to monitor progress by amassing quantitative data regarding the leveraging and use of resources, participation in initiative-sponsored professional development, level of curriculum implementation, and student achievement. The data have been aggregated to report results for the following outcome goals in response to the Government Performance and Results Acts (GPRA):

After three years of support, over 80 percent of schools participating in a systemic initiative will:

- Implement standards-based curriculum in science and mathematics
- Further professional development of the instructional workforce; and
- Improve student achievement on a selected battery of tests.

This division-wide database is designed to collect similar information across the systemic initiatives programs and to make information accessible to multiple parties simultaneously. The collection and analysis components have been linked to expedite the receipt of three types of reports: aggregate reports of core qualitative outcomes, summaries of results by program type, and individual site reports.

Initiatives are also required to submit an annual report via FastLane (the foundation-wide electronic reporting system) and to structure their annual reports to address their performance with regard to the six reform

drivers. In addition to describing their major accomplishments in terms of the drivers, the data are utilized to support strategic plans for the upcoming year.

Reviewing the Effectiveness of Initiatives. Each initiative has to participate in the mid-point review process in year three of their five-year funding period. The mid-point review is a standard procedure in the education directorate for large-scale awards funded over long periods of time. The review involves a panel of outside experts (systemic reform practitioners, researchers, evaluators, and assessment specialists) who examine the accomplishments of the individual initiative and make recommendations about future directions and funding continuation. The findings and recommendations of the panel are shared with the initiative's leadership for mid-course correction, to improve implementation and enhance programmatic success.

The *program effectiveness review (PER)* of each of the systemic initiative (SI) programs was conducted annually for approximately four years. ESR required evidence of each of the initiatives' (RSI, SSI, USI, CPMSA, and UPS) success in changing the K-12 education system for the targeted geographical area and in improving learning for all students. Closely aligned with the mostly qualitative annual report based on the drivers on systemic reform and the GPRA-driven quantitative web based Core Data Elements (CDE) collection, the PER focuses on nine *performance indicators*: student impact; teacher impact; policy change; resource change; management change; data utilization; learning infrastructure change; student performance; and partnership. All or a sample of the principal investigators prepared a data-driven report about the aforementioned nine indicators and made oral presentations to a review team from NSF. The oral presentations highlighted changes in infrastructure and student outcomes, and explained how selected components of their systemic reform implementation strategy improved the quality of K-12 mathematics and science education. This program review practice was instrumental in 1) helping ESR to identify patterns and trends over time in order to make informed educational management decisions and 2) increasing self-reflection and improved data utilization practices at the initiative level.

External Accountability Studies

The Foundation has funded third-party evaluation studies to conceptualize and analyze the implementation processes and measure the outcomes and impacts of the systemic initiative programs. These evaluation studies are needed for objectivity in managing and reporting credible results. Depending on the need for information and the implementation stage of the program, these efforts were funded either as a program evaluation, an evaluative study, or an impact study.

Evaluating Outcomes. The systemic programs have been participating in two types of external evaluation studies, in addition to their local evaluations. First, the Foundation funded through the contract mechanism the traditional *program evaluation* for formative and summative purposes. The program evaluation is usually implemented during the beginning stage of the program as was the case with the SRI's evaluation of the SSI and the COSMOS's current evaluation of the Urban Systemic Program. These studies are usually funded for five years to determine the extent to which the program solicitation goals and requirements have or have not resulted in successful systemic reform.

The second is an *evaluative study* of the program designed to document and present findings within programmatic, national, and international contexts to a variety of audiences. The focus is on successes, pitfalls, and best practices in stimulating, promoting, and sustaining positive changes in mathematics and science teaching and learning outcomes. After the urban and rural systemic programs had been in existence for more than three years, the Foundation supported evaluative studies to advance the knowledge base in two areas: systemic reform (e.g., study of variables that have public/local/political importance and focus on the catalytic/critical events in the reform process; examination of the influence of internal system change versus challenges to the system posed by an outside agency, other sponsors, and partners) and evaluation innovations (e.g., examination of existing data systems as enablers that endow sites with the capacity to monitor reform). Additionally, these studies have focused on all or a sample of sites and considered the following in their approaches to make generalizable inferences about the implementation and evaluation of system reform: diverse team composition, comparative design, and dissemination efforts, as well as other innovative concepts and methods. The evaluative studies had to have a robust design for identifying characteristics or dimensions that allow for comparisons within, between, and among study populations over time for hypothesis testing and/or for discovery about the causes or effects of systemic reform (NSF, 1999).

A key lesson learned from the SSI program evaluation has been the need for both breadth and depth in understanding intended and unintended outcomes. In other words, additional research is needed to enhance the evaluation studies in order to explain fully the outcomes, synthesize and/or reconcile results from different studies, and use the results to inform practice. For example, the SRI International evaluation of the SSI program produced 12 detailed case studies that verified and validated the survey findings, served as the data source for reporting on cross-cutting themes, and offered illustrative examples of the reform strategies and issues for the final synthesis report of key findings and program accomplishments. This was coupled with a deeper investigation of the following topics: professional development, classroom practices, and student achievement (Zucker *et al.*, 1998).

Recently, the Division of Research, Evaluation, and Communication solicited *evaluative research studies* to respond to the dual context (depth and breadth) for understanding systemic reform, recognizing the interactive components of research and evaluation to strengthen education innovations. The evaluative research studies have been designed to increase the theoretical and empirical knowledge base on educational systemic reform in order to enhance the review and assessment of the output and outcomes of NSF's Systemic Initiatives.

Determining Sustainability and Long-term Impacts. Program evaluations and evaluative studies have been conducted in a reasonable timeframe to focus on intermediate outputs and project outcomes. However, to determine the long-term benefits of the reformed education system, the Foundation is funding impact studies after the program has been in existence for about five or more years. The SSI Program is currently in Phase II—a second five-year period to scale up efforts to full implementation of standards-based curriculum and to enhance the capacity of the infrastructure. To complement the program evaluation based on the first five years of statewide systemic reform, the Foundation has funded two impact studies of the SSIs. These impact studies will examine students outcomes and what was institutionalized or sustained after the end of the initial funding period.

COLLECTING AND FINDING EVIDENCE OF EFFECTIVENESS OF SYSTEMIC REFORM IN IMPROVING THE K-12 EDUCATION AND LEARNING

Statewide Systemic Initiatives Program

Program Evaluation. Only one of the systemic initiatives programs administered by ESR has a completed evaluation study. The national evaluation of the SSI program, conducted by SRI International, assessed the SSI as a federal strategy for improving mathematics and science education by documenting educational changes at each stage of the systemic reform process. The conclusion was that the SSI program has provided substantial contributions to standards-based, systemic reform in K-12 mathematics and science education. The key evaluation findings are noted below.

- High-quality, focused intervention by SSIs had some demonstrable, positive classroom impacts. Approximately half of the 22 SSIs completing the 5-year funding period had strong, positive impacts on classroom practices; namely greater use of hands-on work; greater attention to student inquiry; greater use of small group work; improved classroom assessments and more use of science kits, calculators, computers, etc. SSIs with these strong impacts on classroom practice provided teachers with high-quality professional development activities that were both intensive and relatively long-term.
- In a review of seven SSIs, the four with the most credible evidence of positive impacts on student achievement were those with strategies/activities most intensively directed at classrooms and characterized by intensive teacher professional development as well as significant investments in instructional materials/resources. In three of these four cases, the SSIs were able to document modest impacts in closing achievement gaps between students traditionally underserved and their peers. The other three SSIs invested most heavily in activities directed at the alignment of state policies or concentrated on building a state-level infrastructure to support change or building local capacity to reform infrastructure; the effect of these efforts on student learning was indirect, by design.
- The SSI impacts were almost always uneven, affecting some districts, schools, teachers, or students involved in the SSI much more than others. No SSI was able to "go to scale" and intensively affect all teachers statewide. Nonetheless, the SSI program moved classroom practice in directions that are generally considered to be an improvement over past practices.

The study also revealed that there is no "one best way" to reform state education systems. Other findings are reported in Exhibit 1, *Understanding What Works*. The evaluation report indicated that the future challenges include the development/revision of models that involve all teachers in high-quality learning experiences as part of normal operating procedures and closing the gap between groups of students usually underrepresented in mathematics and science and their peers. The study team also reported that future evaluations of large systems need to investigate scale up strategies and share best practices regarding how to go to scale and intensively

Related to the growth of a leadership cadre is the growing knowledge base about "what works" in systemic reform, such as:

- Setting clear and ambitious education standards does, in fact, seem to be an important step in improving education (as the theory of systemic reform suggested it would be).

- Local control need not be a barrier to systemic reform. However, local-control states require different approaches than others.

- Changing whole schools took significant time and resources targeted on a small subset of the system. The results were limited unless the model-school strategy was combined with well-developed strategies to ensure that the lessons from model sites could be used in additional schools.

- Mathematics and science content knowledge is a necessary part of the systemic reform of mathematics and science education. The success of state and federal efforts to improve student achievement in mathematics and science depends on the capacity of individual teachers themselves to thoroughly understand what they must teach students.

- Capacity building is a key to having world-class mathematics and science education, and nearly all SSIs used some of their resources to improve their capacity-building infrastructure. Building infrastructure took different forms in different states: for example, teacher networks, new regional assistance centers, a technology infrastructure, or improved procedures for selecting instructional materials.

- Going to scale with standards-based reform requires states, districts, and schools to work together and takes more than 5 years.

Exhibit 1. Understanding What Works In Statewide Systemic Reform.[1]

affect all teachers and students statewide in mathematics and science (Zucker *et al.*, 1998).

Impact Studies. Currently, two three-year impact studies of SSI are underway. Collaborating with Research Horizon Inc., the evaluation team from the Wisconsin Center for Education Research is studying the impact of the SSI program on student learning, curriculum and policy. A series of studies have been conceptualized, proceeding from investigating what are general relationships between student outcomes and states' participation in the SSI Program to investigating very specific relationships between student

---

[1] From *A Report on the Evaluation of the National Science Foundation's Statewide Systemic Initiatives (SSI) Program* (NSF 98-147) by A. Zucker, P. Sheilds, N. Adelman, T. Cocoran, & M. Goertz, 1991, p. X.

achievement and the SSI as implemented by a state. The two major goals are to assess the impacts of SSIs on student learning using NAEP and state assessment data; and to distill the lessons learned from designing, implementing, evaluating, and supporting SSI. The latter goal will emphasize two critical attributes of systemic reform—strategic decision making, and sustainable reform based on document review and interviews of SSI and state mathematics and science education leaders in 12 SSI states. The focus on strategic decision making will examine how the SSIs took the context of their state education system into account in their design and implementation, how the SSIs positioned their efforts with respect to other education reforms and trends within the state, and how the SSIs have responded to challenges and opportunities over the course of the initiatives. The focus on sustained reform will highlight how the SSIs have prepared for the future, in particular, how the SSIs have maintained important gains they have attained and, how they plan to further those gains in pursuit of longer-term outcomes (Webb, 1998).

In the second multistage, multi-state impact study, evaluators at the University of Minnesota-Twin Cities are incorporating the drivers, and other factors derived from research on school change, into an empirical model to help explain the impact of systemic reform efforts on student achievement. Using a quasi-experimental design, schools in three SSI states with high levels of involvement (treatment) will be contrasted with schools with low levels of involvement (comparison) and with comparison groups in a non-SSI state. Three confirmatory data collections are being employed—surveys, observations, and interviews. Several data analyses will be used including path analysis to better understand the interactive nature of the relationships between and among the SSI drivers and structural equation modeling to better understand the impacts of SSIs on student achievement (Lawrenz, 1998).

Urban Systemic Initiatives Program

Evaluative Studies. Two three-year evaluative studies of the USI program are underway. These studies are in year two of their investigations. One study examines the entire portfolio of USI sites; the other investigates the outcomes of four USIs. Specifically, the evaluative study team at Systemic Research, Inc. is collecting both qualitative and quantitative data that are indicative of systemic change and developing metrics to assess system effectiveness (Kim, 1998). Various statistical methods will be used to develop an inferential causal model linking the results of the rubrics for measuring policy and practice to student outcome data. The analyses includes comparison of progress across 21 USI sites, analysis of the results by duration and intensity of involvement in systemic reform efforts, the measurement of system effectiveness before and after the implementation of systemic initiatives, and the comparison of differential outcomes of USI and non-USI sites. (Systemic Research, Inc., is also conducting a similar evaluative study of the Comprehensive Partnerships for Minority Student Achievement Program.) Preliminary findings of the USI evaluative study include:

- Gate-keeping mathematics and science course enrollment has increased for all students in 17 USI sites.
- The majority of the USI sites have increased their graduation requirements to include more mathematics and science courses. Many of these sites are also requiring specific and more challenging mathematics and science courses. Despite the increase in graduation requirements, the sites are still experiencing a growth in graduation rates.
- In a majority of the USI sites, the overall achievement of all students as measured by the passing rate on the district assessment tests in mathematics and science has improved from baseline year to 1998–99 (Kim, 2001).

The team of evaluators and researchers at the University of South Florida is implementing a multi-site case approach in a series of three investigations, using a structural equation modeling approach to develop an understanding of the causal relationships among the drivers of systemic reform. The first investigation, "The Mathematics and Science Attainment Study," will determine to what extent do student outcomes in general and the attainment differences between underserved students and their peers differ as a function of the duration and intensity of treatment. The second investigation, "The Study of Enacted Curriculum," is focusing on how instruction is changing over time. The third investigation is examining the influence of local USI reform efforts on mathematics and science policies, as well as the extent to which local communities mobilize resources to enable students to achieve high standards in mathematics and science (Borman, 1998).

Rural Systemic Initiatives Program

Evaluative Study. The evaluation team from the University of Western Michigan is conducting an evaluative study of the RSI Program with a sample of six RSI sites. The objectives are: 1) to develop a system of indicators around each of the identified six drivers of educational system reform; 2) to determine the perceived relative importance and value of each of the drivers and indicators for reform in RSI schools in selected communities; 3) to determine the status of innovation/reform within selected communities with respect to factors thought to support or serve as barriers to innovation and education reform; and 4) to determine the ways and the extent to which the perceived importance and value of drivers and the characteristics of the community impact on systemic reform efforts and student achievement in mathematics, science and technology (Horn, 1998). The project objectives are being investigated using a mixed methods approach, including case studies, surveys, Delphi techniques, interviews, focus groups, secondary analysis of achievement test data, and on-site observations. An important outcome will be the empirical validation of the drivers of educational system reform. Some initial findings are:

- Rural schools are very effective in converging monies (local and grant) to support reform efforts in science and mathematics.
- Mathematics and science curricula are being aligned with standards (mostly state standards), and are being driven by mandated statewide testing.
- The most effective elements of the RSIs seem to be assistance in aligning curricula to established standards; professional development for large numbers of qualified and unqualified mathematics and science teachers; financial assistance to support a teacher partner to provide local assistance and leadership; curriculum audits, and assistance in moving mathematics and science improvement to a place of high priority (Horn, 2001).

Urban Systemic Program

Program Evaluation. The COSMOS evaluation team is employing a case study approach to track processes and outcomes over time for the USP (Yin, 2001). Using cross-site methodology, the data from multiple USP sites will be analyzed as if they were data from multiple experiments. The framework for interpreting cross-site pattern in USP is based on replication logic, determining the extent to which patterns in USP sites emulate the program logic models. Additional analyses are also planned, including comparisons that account for different levels of intervention and the assessment of major rival conditions that are potential threats to the USP efforts. The first year has focused on the identification of early signs of "systemic-ness." Some preliminary examples of progress across the sites include:

- Completion of the district-wide "visioning" process that linked the vision for reform with the distribution of curriculum guides and other classroom materials
- The linking of assessment with other components by reviewing test scores not only for students' performance but also for the extent to which a standards-based curriculum had in fact been implemented and the professional development levels of the teachers
- Partnerships with institutions of higher education (IHEs) to align preservice education with reform curricula, including having IHEs' upper division students serve as tutors in K-12 classrooms
- The linking of the "broad-based support" and "curriculum" components by initiating parent centers within reforming schools, to identify and train "lead" parents to provide other parents with information on standards-based education and recommending parenting activities to reinforce mathematics and science learning

The study team will also monitor barriers to progress, such as changes in accountability polices which may impact comparison of achievement data, a site's scale up strategies, districtwide budget freezes, and changes in district leadership.

EVALUATION ISSUES IN SYSTEMIC REFORM

A number of complex issues are being addressed in the evaluation portfolio of ESR's systemic reform programs. The following discussion will focus on the issues that have emerged and how they are being addressed in the NSF-funded third party evaluation and/or evaluative research studies.

Conceptualizing Systemic Reform

A major activity in assessing progress and determining the attainment of outcomes is the development of a conceptual framework of the logical flow of events leading to the desired changes. Several conceptual and/or logic models have been developed by evaluators to capture the concept of systemic reform and guide data collection and analysis. Examples include those developed by SRI International; Horizon Research, Inc.; and the University of South Florida (See Exhibits 2–4).

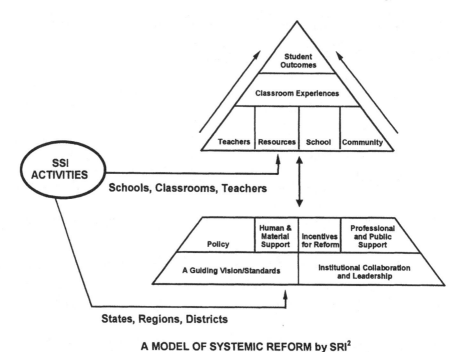

A MODEL OF SYSTEMIC REFORM by SRI[2]

Exhibit 2. The Model of Systemic Reform for the SSI Program Evaluation.[2]

A critical dimension in the reform models is the relationship among the components to depict the necessary alignment to produce comprehensive change and continue to change positively over time for all students. In addition to a logic model for the course of events at an USP site (Exhibit 5), COSMOS Corporation has developed a second program logic model (Exhibit 6) to depict the dynamic and multi-faceted nature of systemic reform as it is progressing. Due to the range and diversity of organizational entities involved in systemic change, this alternative model is needed to counter the finding that early data might be misinterpreted as not corroborating the linear logic models. More importantly, the second model is a

---

[2] Taken from *A Report on the Evaluation of the National Science Foundation's Statewide Systemic Initiatives (SSI) Program (NSF 98-147)* by Andrew A. Zucker, SRI International; Patrick M. Shields, SRI International; Nancy E. Adelman, SRI International; Thomas B. Corcoran, Consortium for Policy Research in Education; and Margaret E. Goertz, Consortium for Policy Research in Education, p. 4.

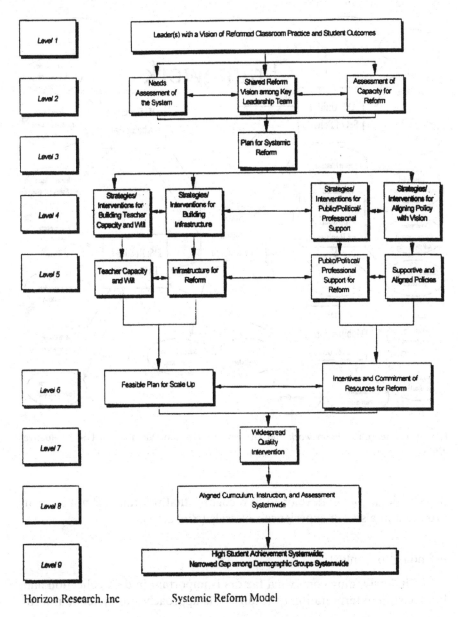

Horizon Research. Inc   Systemic Reform Model

Exhibit 3. Model of Systemic Reform.[3]

---

[3] Taken from "*Study of the Impact of State Systemic Initiatives*," an NSF funded project submitted by Norman Webb, Wisconsin Center for Education Research and Iris Weiss of Horizon Research Inc. (REC-9874171), p. 14.

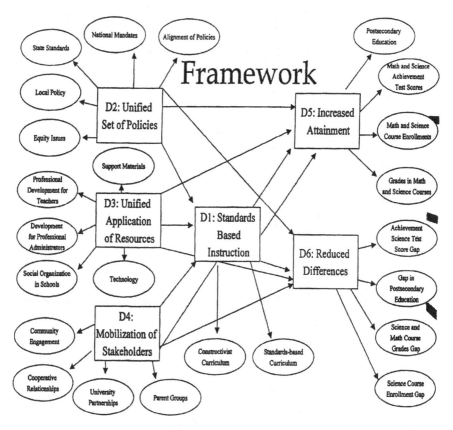

Exhibit 4. Theoretical Framework and Drivers and Indicator Variables in USI Evaluative Study.[4]

dynamic logic model developed to recognize that reforming is not a linear process and also can experience reversals (Yin, 2001).

Attending to Context

Understanding contextual factors is important in data collection and it affects the interpretation of results. SSIs approaches to systemic reform

[4] Taken from "*Assessing the Impact of the National Science Foundation's Urban Systemic Initiative*", an NSF funded project submitted by Kathryn Borman of the University of South Florida (REC-9874246), p. 15.

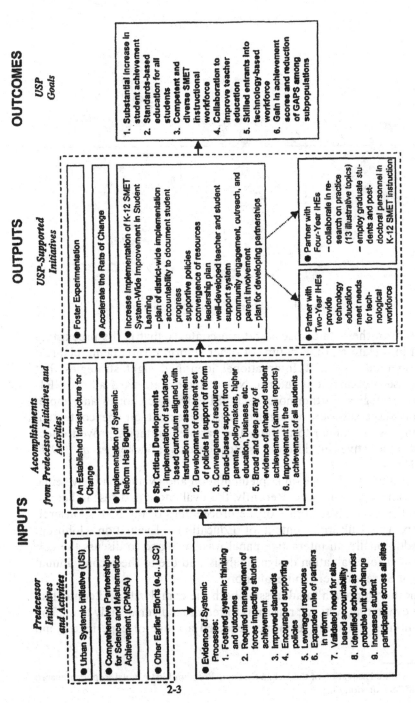

Exhibit 5. Logic Model for Course of Events at an Urban Systemic Program (USP) Site.[5]

[5] Taken from *Cross-Site Evaluation of the Urban Systemic Program First Annual Report* by R. Yin, A. Noboa-Rios, D. Davis, I. Castillo, & R. MacTurk, 2001, pp. 2–3.

INPUTS

Predecessor Initiatives and Activities

- Urban Systemic Initiative (USI)
- Comprehensive Partnerships for Science and Mathematics Achievement (CPMSA)
- Other Earlier Efforts (e.g., LSC)

- Evidence of Systemic Processes:
  1. Fostered systemic thinking and outcomes
  2. Required management of forces impacting student achievement
  3. Improved standards
  4. Encouraged supporting policies
  5. Leveraged resources
  6. Expanded role of partners in reform
  7. Validated need for site-based accountability
  8. Identified school as most probable unit of change
  9. Increased student participation across all sites

2-3

Accomplishments from Predecessor Initiatives and Activities

- An Established Infrastructure for Change
- Implementation of Systemic Reform Has Begun
- Six Critical Developments
  1. Implementation of standards-based curriculum aligned with instruction and assessment
  2. Development of coherent set of policies in support of reform
  3. Convergence of resources
  4. Broad-based support from parents, policymakers, higher education, business, etc.
  5. Broad and deep array of evidence of enhanced student achievement (annual reports)
  6. Improvement in the achievement of all students

OUTPUTS

USP-Supported Initiatives

- Foster Experimentation
- Accelerate the Rate of Change
- Increase Implementation of K-12 SMET System-Wide Improvement in Student Learning
  - plan of district-wide implementation
  - accountability to document student progress
  - supportive policies
  - convergence of resources
  - leadership plan
  - well-developed teacher and student support system
  - community engagement, outreach, and parent involvement
  - plan for developing partnerships

- Partner with Two-Year IHEs
  - provide technology education
  - meet needs for technological workforce

- Partner with Four-Year IHEs
  - collaborate in research on practice (13 illustrative topics)
  - employ graduate students and postdoctoral personnel in K-12 SMET instruction

OUTCOMES

USP Goals

1. Substantial increase in student achievement
2. Standards-based education for all students
3. Competent and diverse SMET instructional workforce
4. Collaboration to improve teacher education
5. Skilled entrants into technology-based workforce
6. Gain in achievement scores and reduction of GAPS among subpopulations

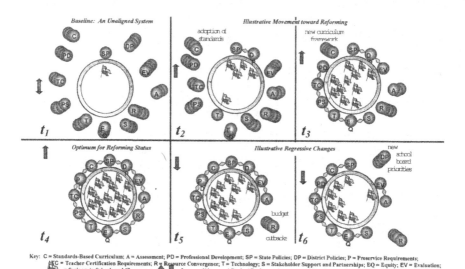

Key: C = Standards-Based Curriculum; A = Assessment; PD = Professional Development; SP = State Policies; DP = District Policies; P = Preservice Requirements; TC = Teacher Certification Requirements; R = Resource Convergence; T = Technology; S = Stakeholder Support and Partnerships; EQ = Equity; EV = Evaluation; = Scale up in Schools and Classrooms; = Increased/decreased Student Performance; = Aligned

Exhibit 6. Hypothetical States of a Reforming System.[6]

were based on their state specific reform contexts, that is the preexisting situation in a state and the nature of other reform efforts being planned or implemented. The SRI International identified eight strategies that the SSIs used to carry out reform. Not a single state used all the SRI-identified strategies. Furthermore, each SSI's strategic choices and implementation successes were depended to some extent on the contextual factors that were unique to a state. The various strategies proved effective and ineffective in different state contexts (Zucker *et al.*, 1998; Westat*McKenzie, 1998).

The cultural influence on implementation and evaluation of systemic reform is an important issue. Recently, the evaluators and researchers of these studies were given copies of the proceedings of the NSF-sponsored workshop, The Cultural Context of Educational Evaluation (Johnson, 2001). Additionally, an interactive panel was convened to discuss the cultural context for evaluation design, analysis, and use at the second annual meeting of SI researchers and evaluators.

Another approach to addressing both the complexity and contextual issues in evaluating systemic reform is to require a diverse study team. The composition of the study team has reflected a range of experiences and

---

[6] From Cross-Site Evaluation of the Urban Systemic Program First Annual Report prepared by COSMOS Corporation.

perspectives consisting of database developers, assessment experts, researchers, systemic reform practitioners, education/policy writers, and evaluators. Additionally, these studies have given serious consideration to diversity in establishing external advisory groups to guide the direction of studies. Multi-disciplinary diversity has been encouraged, also. For example, an assessment study of student achievement in urban settings is staffed with a team consisting of an evaluator, an educational researcher, a sociologist, an economist, an educational psychologist, and a political scientist (Clune, 1999).

Focusing on Systemic Reform Attributes

Evaluating components in isolation does not reflect the nature or complexity of system change. The program evaluation of USP is tracking signs of *systemicness* while viewing different component working in tandem. The COSMOS evaluation team has stated that this is a more cogent approach, reflecting that the essence of reform is to attain alignment among a system's components and not just to advance them individually (Yin, 2001). Similarly, the impact study of the SSI program is developing data collection tools and metrics that cut across the systemic reform drivers to report on a set of systemic reform attributes: alignment, saturation, equity, linkages, capacity, and sustainability (Webb, 1998).

Reducing Data Collection Burden

The COSMOS program evaluation team is conducting a cross-site evaluation of the USP while the individual performance of each USP site is assessed by the site's own local evaluator. Many of the same type of data are relevant to both the program evaluation and local evaluations. COSMOS is establishing a collegial arrangement with local evaluators who have agreed to share data and the results of their local evalutions, thereby maximizing the benefits of both types of evaluation efforts, (Yin, 2001).

NSF has had some success in encouraging and sponsoring efforts to network the evaluators and researchers of systemic reform in order to avoid duplication of data collection. Systemic Research, Inc., for example, hosts a web site (www.siurbanstudy.org) to facilitate communication and collaboration among the researchers and evaluators of urban systemic reform (Kim, 1998).

Attention to respondent burden has been an important factor in the development and revision of the web-based Core Data Elements (CDE) collection. The CDE web-based system has reduced respondent burden by including such features as automated tabulation; data entry using pick lists, check boxes, and option menus; data verification with error messages for on-line-corrections; standard reports; and pre-defined tables, charts, and graphs. Additionally, the evaluators and the researchers of the systemic reform studies have been using CDE in streamlining or augmenting their own data collection efforts.

Identifying the Unit of Change/Analysis

The sites vary in the unit of change which helps to inform sampling and analysis decisions. The studies have been sensitive to ESR's concerns that the unit of analysis is reflective of the unit of change. One approach is identifying the unit of change as the entity in the system that must be restructured as a necessary condition for improving teaching and learning (e.g., classrooms). Another approach is focusing on the administrative unit of the system that must be changed in order to promote accountability, institutionalization, and sustainability of systemic reform beyond the duration of NSF funding (e.g., school districts).

In several studies the unit of analysis will include both change elements in order to track the effectiveness of implementing specific drivers as well as the system-wide results of overall reform activities. For example, in the evaluative study of the USIs, the district is used as the unit of change for the cohort analysis and reporting of successful systemic practices and student accomplishments. The same study has identified the teacher as a unit of analysis to examine changes in classroom practices in mathematics and science (Kim, 1998).

Using Appropriate Comparisons or Controls

With or without NSF funding, districts and states across the nation are pursuing standards-based reform, making it difficult to find sites totally absent of restructuring mathematics and science education. The studies are aiming to control for common demographic variables and the amount of SI effort that had been expended in the district. However, the scale up aspect of systemic reform proposes a continuing challenge. That is, if the reform is becoming more systemic over time, there is a gradual loss or blurring of comparison/control groups (Kahle, 1999).

Expecting Improved Performance

Most of these accountability studies have pointed out that improved student achievement often does not show up until after the students have been exposed to the changes over a number of years. The SRI International,

COSMOS Corporation and Systemic Research, Inc. evaluation teams have emphasized that the precise length of time for significant results related to the outcome drivers is difficult to estimate because of the complexity of systemic reform and the time needed to implement and sustain the critical process drivers. The full effects of the reform will only be realized when all students have been fully exposed to the reformed system throughout all their grade levels. Despite this fact, the current evaluation studies have noted that improved student performance should nevertheless start to appear before the end of a site's five-year period as a systemic initiative.

Addressing Concerns about Measures and Methods

Criticisms of some traditional methods are being addressed in current program evaluations and evaluative research. One issue is the need to develop reliable methods to evaluate change in classroom instruction and teacher content knowledge (Blank, 2000). The concern is whether teachers' self reports on surveys accurately describe the extent to which they are implementing the mathematics standards in their classroom when compared to classroom observations of teaching, curriculum, assessment, and group practices. The question remains regarding which is the most accurate method for measuring the degree of implementation of mathematics standards in a classroom. Additionally, there is a need for comparable measures of professional development and its impact on teaching.

Researchers at Rutgers University have been funded to develop valid methods to assess teaching practice and its change over time (Firestone, 1999). They are developing self-report dimensions that are triangulated using observational data. Classroom observations will be summarized in a format that provides a holistic view of the class session but also focuses on the dimensions built into the survey items—i.e., Active Learning, Multiple Representation, Collaborative Learning, Deep Learning, and Classroom Assessment. The study team will use confirmatory factor analysis and more conventional statistical procedures to develop a set of measures of instructional practice that have reasonable reliabilities and other measurement properties. These measures will then be further validated through correlation with direct observation and artifact data.

Opinions about the use of qualitative evidence to evaluate the effectiveness of systemic reform have been mixed. Several evaluators and researchers of sytemic reform, who attended a meeting at the Foundation in 1999, shared insights on how to improve the credibility and persuasiveness of case studies. The following ideas were offered to give attention to the needs of the audience who are less accustomed to working with/relying on qualitative data reporting:

- Include not only process information but also outcome information and consider placing the outcome information up front, followed by narrative about process.
- Explicitly clarify how cases and sub-units were selected. Include presentation of contrasting perspectives, noting common opinions and unique perspectives.
- Include work and numerical tables in the case study document, as well as vignettes to illustrate quantitative data.
- Use diagrams to show empirically derived process data (e.g., show the logic model as tested, not just as anticipated).
- Most importantly, triangulate inductive reasoning with evidence from other sources (Darke, 1999).

Confronting the Challenges in Analyzing Student Achievement Data

The program evaluation of SSI identified several factors that impacted the investigation of the relation between the reform and student achievement: lack of common assessment instruments, difficulty identifying level of engagement in SSI, and lack of clearly defined comparison groups, inability to disaggregate assessment information into scales or by items (Laguarda, 1998; Laguarda, Breckenridge, Hightower & Adelman, 1994). Additionally, it has been emphasized that improvement in performance scores can be achieved without the reduction of the "gap" between majority and historically underrepresented students. It is, therefore, more difficult to eliminate the "gap" than it is to raise a school's test scores. Another point is that adequate science assessments with known reliability and validity, adequate controls over demographic characteristics of students, and adequate content coverage are not yet available.

Since sites use different measures to assess student achievement, an equating process is needed to compare data from different assessment instruments. One response to this challenge is that of Wisconsin's SSI impact study. The study team will use item response theory to equate the assessment instruments. Data from national and state assessments will be transformed into comparable achievement variables such as mathematics computation, mathematics problem solving, and science concepts (Webb, 1998).

Attributing Results to the NSF Investment

Attributing system change and its influence on improved student achievement to systemic reform efforts requires non-traditional methods (Chubin, 1997; Frechtling, 1995). The evaluators have underscored that the conditions for attribution include random assignment, a control group, and common achievement measures across sites but these conditions are lacking in the implementation and study of systemic initiatives. Also, the fact that NSF support serves as a catalyst to advance systemwide reform increases the difficulty in delineating the influence of the SIs from other reform efforts.

In the absence of conducting a true experiment, COSMOS is naming and testing rival explanations to increase the certainty of causal attributions. The more the named rivals are plausible, important, and rejected, the stronger the causal attribution to the target systemic reform effort. The Wisconsin study team has placed strong emphasis on the need to validate and triangulate information to eliminate possible alternative hypotheses. The team will construct a chain of evidence about how the quality and saturation of SSI implementation impacts student achievement, and how the interactions of SSI with other reform efforts relate to improvements in student achievement.

Another approach is that of the University of Minnesota in which one of the sample sites will serve as a "replication study" for validation and verification purposes. In the evaluative study of USIs conducted by the University of South Florida, the sampling strategy will be one of selecting matched pairs of schools (each pair consisting of a "high reform" and a "low reform" school). The pairs are matched on the most salient extraneous variables that are potentially related to the drivers. By controlling for extraneous variables and using structural equation modeling, the study team will be able to examine the unique contributions of systemic reform to student achievement outcomes.

The principal investigator of the Bridging the Gap study noted that the design did not propose to attribute causality; however, the use of multiple data sources to examine similar trends and the descriptive and inferential analyses do suggest that the positive finding favoring the systemic reform effort is more than a chance phenomenon (Kahle; 1999). Within the next two to three years this statement may be applied to several of the evaluative research studies. Generally, the evaluation and research studies are addressing issues of attribution theoretically and empirically with sample selection and statistical analyses.

Improving Data Utilization

Internal assessment activities and external evaluations have found that states and districts need help in developing their capacity to monitor systemic reform. The SSI summary report by staff at the Council of Chief State School Officers (CCSSO) pointed out that data are not accessible at different levels of the system in most states, and data reports are usually not formatted and presented in ways that are useful or understandable to educators, administrators, or policymakers (Blank, 2000). Systemic change and the evaluation of systemic reform can be advanced by 1) presenting data with a goal to increase applications of data for instruction, 2) understanding and using appropriate statistics, 3) disaggregating data, and 4) presenting results in understandable format that will allow stakeholders to answer "what does the data tell me?"

Several accountability studies will link their evaluation data to state information systems. The study group at Wisconsin, for example, will develop an information network containing data from each of the states. The information network will have the capability for new information to be added, such as additional assessment data, and for electronic inquiries to be made. Inquiries about descriptive data, such as what had been the change in mathematics achievement over time, and about the relation among variables organized in the large categories of student outcomes, process indicators, and systemic reform attributes will be possible. The information networks will be a tool that people will use to know more about their education system (Clune, 1999). A research team at the University of Arkansas is developing a step-by step approach for data collection, inclusion of appropriate analyses, and enhanced reporting of achievement results for statewide initiatives to improve the evaluation of relative gains in performance that can be credited to system reform. This product will be made accessible electronically to educators, researchers, and evaluators (Mulvenon, 1999).

Needing More Than One Type of Study

In June 2000, the Council of Chief State School Officers (CCSSO) convened a meeting of researchers, evaluators, and state program leaders of the SSI program to review the major research and evaluation findings and to discuss lessons from the states' experiences with SSI (Blank, 2000). Exhibit 7 provides some of the lessons learned. With the growing knowledge base about "what works" and "lessons learned", evaluators and researchers were in agreement that much remains to be learned about the practice and theory of systemic reform implementation and evaluation in education. The nature of systemic reform is complex and no one evaluation study has sufficient resources and time to fully investigate the breath and depth of all the components of restructuring education systems. Systemic reform involves the simultaneous restructuring of many components of the education system in order to improve simultaneously the academic performance of all students at all levels of the K-12 system,

1. Setting a common vision for improving mathematics and science education provided a basis for systemic planning, consistent models and strategies for improvement, and criteria for evaluating progress.

2. Leaders of systemic reform must have credibility and capacity in order to develop connections with bases of power and support, including partner organizations and policy-makers.

3. A critical role for state is in the long-range planning of human and fiscal resources needed for systemic change, as well as the development of shorter-term strategies and leveraging of available funds. Effective SSIs developed important partnerships with education policymakers, leaders of science and math professions, higher education, and business and industry; these partners were critical for program vision, support for goals, and leveraging resources.

4. Challenging content standards, clearly written, are an important component of systemic improvement, and successful state initiatives ensured that assessments, curriculum, and instruction were aligned with standards.

5. The professional development models in effective SSIs focused on math and science content knowledge and active learning strategies, and the experiences of teachers were typically coherent and sustained.

6. Key components for sustainability identified during the SSI program are: maintaining quality control, building incrementally on prior reform, leveraging funds, and infrastructure.

7. Lack of alignment between existing assessment instruments and state standards and goals for improving learning impedes state and national efforts to evaluate the program's real impact, and lack of integrated state data systems inhibits efforts to analyze equity issues in program impact.

8. Several crucial elements of long-term systemic improvement remain on the agenda for math and science in most state, including
   • Establishing an assessment system consistent with content standards
   • Passing state legislation and policies consistent with systemic improvement goals
   • Consensus building with districts, school, and the public about the vision for mathematics and science reform

Exhibit 7. Lessons Learned Concerning State Systemic Reform of Science and Mathematics Education.[7]

---

[7] Adapted from "*Summary of Findings from SSI and Recommendations for NSF's Role with States: How NSF Can Encourage State Leadership in Improvement of Science and Mathematics Education*," Rolf Blank, CCSSO (2000, draft).

as indicated by the four process and two outcome drivers of systemic reform.

Since 1999, the Foundation has funded approximately 12 evaluative research studies to generate new and/or increased knowledge about the systemic reform drivers and the evaluation of large-scale systemic reform. These studies will complement the accountability studies designed to gauge success in attaining program outcomes by focusing on the depth or the breadth of specific implementation/impact areas of systemic reform. Some of the evaluative research studies are highlighted below, organized by the systemic reform drivers discussed earlier.

Classroom Driver: In a study of state initiatives, researchers at CCSSO are measuring differences in instructional practice and curriculum content among teachers and schools to determine if there are differences in mathematics and science teaching that are related to state policy initiatives and state standards (Blank, 1999). A research team at Education Development Center is investigating the role of teacher leadership for transforming classroom practices and as an overall strategy for achieving reform in science and mathematics education (Miller, 1999).

Policy Driver: Clune's study of SSI (1999) stressed the importance of guiding policies or supportive authority from the state and financial support as being among the key characteristics of successful reform. Clune has emphasized that systemic reform should be represented as a continuous casual sequence beginning with the baseline of prior policy:

> PP →
> SR → SP → SC → SA
> (where PP = prior policy, SR = systemic reform, SP = systemic policy, SC = systemic curriculum, and SA = student achievement corresponding to the curriculum)

Resource Driver: The Urban Institute is examining why and how school districts reallocate resources as a result of reform (Hannaway, 1999). Advancing the understanding of NSF's role as catalytic in leveraging and coordinating resources, COSMOS is studying active SSIs to determine how much "reforming" is occurring and what has been the role of NSF's SSI in producing "reforming" (Yin, 1999).

Stakeholder/Community Support Driver: Another research team at Education Development Center is focusing on stakeholders in rural science education who participate in NSF's RSI program. The study is expected to enhance our understanding about rural communities access to and engagement in the RSI Program and will offer recommendations for reaching increased numbers of rural communities with reform efforts (Century, 2000).

Attainment Driver: A study at the University of Maine is exploring the use of national and state assessment data to understand the performance of two SSI states. The study team will examine the methodological challenges posed by multi-level, multi-measure, time-series student assessment data in obtaining value-added school performance indicators and determining if the national and state assessments produce consistent results on the proficiency of differing groups of students. The study is expected to help educators and evaluators become more aware of the technical constraints on current national and state assessment, any biases that may have been a part of the assessment processes and results, the alternatives available, and the consequences of assessment reform (Lee, 1999). The goal of a study focusing on three CPMSA sites is to test the systemic reform theory by estimating the quantitative relationship between individual elements of the process drivers and the change in student achievement (Cavalluzzo, 1999).

In 2000, EHR solicited and funded five studies to evaluate large-scale student outcome datasets from the SI awardees to clarify and extract in greater detail the attribution of whole system progress to the SI investment. A study being conducted at the Wisconsin Center for Education Research will answer questions like: What analytic statistical models best fit the data linking the systemic initiatives to student achievement and how can these be demonstrated in prototype analyses? What lessons can be communicated to the field about the kind of database design and analysis that is more, or less, useful in evaluating and understanding systemic reform (Clune & Webb, 1999). Measurement specialists at the University of North Carolina are using meta-analyses to examine the relationship between SI involvement and student achievement within and across initiatives (Harman, 1999).

Equity Driver: The Urban Institute is collaborating with Campbell-Kibler Associates, Inc, in a three-year study focusing on identifying and describing characteristics of highly effective urban schools serving students most in need (Clewell, 1999). One of the questions in this study is whether all subpopulations being served by the USIs are benefiting equally from systemic reform. The Bridging the Gap study at Miami University has been focusing on how systemic reform works in schools that are at different stages of readiness for reform to understand performance disparities and to identify best practices for accelerating the achievement gains of underrepresented populations in mathematics and science (Kahle, 1999).

FUTURE EVALUATION CHALLENGES AND DIRECTIONS

Borman and colleagues have emphasized that commonalties and differences among the SI sites create both opportunities and problems in the identification of indicators and in making sampling, measurement, and aggregation decisions (Borman, 1998; Kromrey, 1999). The evaluation designs discussed in this chapter have been reviewed internally and externally. They were found to be appropriate to the context of the systemic initiatives programs, according to the report of the recent Committee of Visitors (COV) for the Evaluation Program at NSF (Evaluation COV Report, 2000). Multiple strategies based upon the requirements of the program and conceptual and methodological challenges of evaluating systemic reform are being used. The mixed-method of using both quantitative and qualitative indicators, use of multiple data sources, and the triangulation of results are common features of the evaluation designs. In addition to advancing the knowledge base on systemic reform, each study strives to develop and/or improve credible innovative evaluation methods.

On the other hand, Webb has stated that systemic change is fraught with tensions between deep learning and skill learning, between raising academic standards and serving all students equitably, and between public goals and educators' goals. These issues will continue to generate enormous evaluation challenges and innovative practices in documenting and judging: how the system components are aligned in purpose and practice; how a system initiative has "gone to scale;" how a system's design can sustain continuing growth in student achievement; how the system has become more equitable for all students; and how improved student learning of important science, mathematics, engineering, and technology permeates the K-16 system and beyond (Webb, 1998).

There continue to be several other areas in need of more evaluation attention. For example, the study team at the University of South Florida is beginning to examine the role of institutional and individual leadership in focusing, coordinating, and maximizing resources that impact student achievement and related outcomes, as the K-12 emphasis on partnerships increase. More focused investigations are needed to understand the various scale up strategies and how reform plans are impacted by targeting the schools most ready versus schools with the greatest need. There is a need to create and disseminate empirically validated instruments to enhance sites' assessments of their progress toward a fully reformed system and appropriately claim the specific contributions of the systemic initiative. Evaluators and researchers of these NSF-funded studies are expected to employ and make available to the field valid instruments and analytical techniques for measuring the systemic reform attributes (e.g., capacity, alignment, linkages, saturation, equity, and sustainability) as well as "systemicness" (interaction of two of more components of reform and how they are leveraged as primary forces in enabling the system toward high achievement for all).

ESR and other divisions in the education directorate are strongly interested in evaluation data to inform equity issues. One of the priority concerns is reducing the correlation between student demographics characteristics and science achievement to zero. The reduction in the achievement differences across racial and ethnic groups and between gender groups is a noteworthy accomplishment for the systemic reform efforts. However, the long-term goal in serving all students with an equitable education system is to use assessment results to help systems change such that achievement outcomes can no longer be predicted based on gender, ethnicity, class, disability, or language group (Anderson *et al.*, 1998).

The goal of K-12 systemic reform in mathematics and science education was, and continues to be, to move from independently devised reform efforts to systemwide, ongoing restructuring involving the coordinated improvements of such elements as professional development, instructional materials, the assessment of student proficiency, education policy, and the use of financial and other resources. The overarching aim of the evaluation portfolio is to corroborate or refute the guiding premise that all children can learn rigorous science, mathematics, and technology. Evaluation results must continue to inform the sites and the larger community of the process in which education systems evolve from an infrastructure focused on delivering instruction to an infrastructure focused on high achievement by students. Williams has described the reform pursuit as one in which the stakeholders must pursue a learning infrastructure rather than an instructional infrastructure (William, 1996). Therefore, multiple data collection methods and descriptive and inferential statistics must be employed to make valid judgments about a (reformed) learning infrastructure where:

1. Instruction challenges all students while focusing on their particular strengths and needs.
2. Standard guides high-level science and mathematics instruction without hampering creativity of teachers, specialists, or principals.
3. Curriculum is research-based, aligned with national standards, applicable to state and local frameworks, and sensitive to the learning needs and styles of diverse groups of students.
4. Assessment provides useful information to teachers about individual student learning and allows the system to account to the public about progress.

5. Professional development is long-term, instructionally ori-
    ented, content-intensive, and responsive to the instructional
    needs of students with a wide variety of backgrounds and
    learning styles.
6. Policies mandate the delivery of challenging science and math-
    ematics content to all students, support the above mentioned
    instruction, and ensure equitable access and high expectations
    for learning.
7. Leadership and partnerships manage and tap resources to
    promote the educational achievement of all children and keep
    the system focused on the necessity for children from diverse
    backgrounds to be prepared for a lifetime of learning and/or
    use of advanced science and mathematics.

Chubin has emphasized that systemic evaluation, like longitudinal
analysis, is a longer-term proposition because it must measure what takes
time—change in capacity and infrastructure that produce greater positive
effects on the whole system. Measuring systemic reform must entail more
than measuring the performance of the system's components. Thus, the
implication is that just as systemic change demands new strategies, systemic
evaluation demands new metrics and innovative methods of assessing
system capacity, infrastructure and performance (Chubin, 1997).

Findings from internal assessment activities and preliminary results of
external evaluative and accountability studies continue to corroborate the
finding of the national SSI program evaluation that the nation's children
are benefiting from improved classroom learning environments, better
quality standards-based curricula, more high skilled teachers in both
content and pedagogy, and a deeper leadership pool of strategic thinkers of
systemic reform. Thus, it is the role of evaluation to provide credible evidence to inform funding decisions and to generate knowledge that will
advance the practice of restructuring the K-12 education systems.

REFERENCES

*Proposal funded as part of the evaluation of ESR's portfolio of systemic initiatives programs.
Anderson, B., Campbell, P., George, Y., Jolly, E., Kahle, J., Kreinberg, N., Lopez-Ferrao, J., and
        Taylor, G. (1998). *Infusing equity in systemic refrom: An implementation scheme.*
        Westat*McKenzie Consortium, Washington, DC.
Blank, R. (1999, December). Study of state initiative in mathematics and science using surveys
        of enacted Curriculum. Presentation at the Conference of SI Research/Evaluation/
        Impact Studies, Arlington, VA.
Blank, R. (2000). *Summary of findings from SSI and recommendations for NSF's role with
        states: How NSF can encourage state leadership in improvement of science and mathe-
        matics education.* Council of Chief State School Officers, Washington, DC.

*Borman, K. (1998). *Assessing the impact of the National Science Foundation's Urban Systemic Initiative.* Unpublished manuscript, University of South Florida, Tampa, FL.

*Cavalluzzo, L. (1999). *An empirical test of the theory of systemic reform.* Unpublished manuscript, CNA Corporation, Alexandria, VA.

*Century, J. (2000). *Science education and rural settings. Understanding access to and engagement in the Rural Systemic Initiatives.* Unpublished manuscript, Education Development Center, Inc., Newton, MA.

Chubin, D. (1997). Systemic evaluation and evidence of education reform. In Bartels, D. and Sandler, J. (Eds.), *Implementing science education reform: Are we making an impact?* American Association for the Advancement of Science. Washington, DC.

*Clewell, B. (1999). *A study of highly effective USI schools.* Unpublished manuscript, Urban Institute, Washington, DC.

Clune, W. (1999, December). *Policy and evalaution.* Paper presentation at the Conference of SI Research/Evaluation/Impact Studies, Arlington, VA.

*Clune, W., and Webb, N. (1999). *Systemic Initiatives: Student Achievement Analysis Study.* Unpublished manuscript, University of Wisconsin—Madison, WI.

Darke, K. (1999). *Summary of breakout session on case study research.* Presentation at the Conference of SI Research/Evaluation/Impact Studies, Arlington, VA.

Evalaution Committe of Vistors Report (2000). Unpublished manuscript, National Science Foundation. Arlington, VA .

*Firestone, W. (1999). *Supporting and measuring high quality instructional practice.* Unpublished manuscript, Rutgers Univeristy, New Brunswick, NJ.

Frechtling, J. (Ed.). (1995). *FOOTPRINTS: Strategies for non-traditional program evaluation* (NSF 95-41). National Science Foundation. Arlington, VA.

*Hannaway, J. (1999). *Strategic resource allocation: Adaptation, capacity, and constraint.* Unpublished manuscript, Urban Institute, Washington, DC.

*Harman, P. (1999). *ESR student achievement data set analysis.* Unpublished manuscript, University of North Carolina at Greensboro, Greensboro, NC.

*Horn, J. (1998). *Rural Systemic Initiative evalautive study.* Unpublished manuscript, Western Michigan University, Kalamazoo, MI.

Horn, J. (2001, January). *Rural Systemic Initiatives evaluation study.* Paper presentation at Second Conference of SI Research/Evaluative/Impact/Assessment Studies, Arlington, VA.

Johnson, E. (2001). *The cultural context for educational evaluation: The role of minority evaluation professionals* (NSF 01-43). National Science Foundation. Arlington, VA.

Kahle, J. (1999, January). *Bridging the gap: Equality in systemic reform.* Presentation at the Performance Effective Review of K-12 Assessment Activities, Arlington, VA.

*Kim, J. (1998). *How reform works: An evaluative study of NSF's Urban Systemic Initiatives.* Unpublished manuscript, Systemic Research, Inc., Boston, MA.

Kim, J. (2001, January). *Preliminary findings from "How reform works: An evaluative study of NSF's Urban Systemic Initiatives".* Paper presentation at the Conference of SI Research/Evaluative/Impact/Assessment Studies, Arlington, VA.

*Kromrey, J. (1999). *Analyzing student achievement in Statewide, Urban, and Rural Systemic Initiatives.* Unpublished manuscript, University of South Florida, Tampa, FL.

Laguarda, K. G., Breckenridge, J. S., and Hightower, A. M. (1994). *Assessment programs in the Statewide Systemic Initiatives (SSI) states: Using student achievement data to evaluate the SSI.* Policy Studies Associates. Washington, DC.

Laguarda, K. G. (1998). *Assessing the SSI's impacts on student achievement: An imperfect science.* SRI International. Menlo Park, CA.

*Lawrenz, F. (1998). *SSI impact study.* Unpublished manuscript, University of Minnesota, Minneapolis, MN.

*Lee, J. (1999). *Exploring data and methods to assess and understand the performance of SSI states: Learning from the cases of Kentucky and Maine.* Unpublished manuscript, University of Maine, Orono, ME.

*Miller, B. (1999). *Teacher leadership for systemic reform.* Unpublished manuscript, Education Development Center, Inc., Boston, MA.

*Mulvenon, S. (2000). *Assessing the Impact of Systemic Initiatives Program: Modeling Best Practices of Data Collection, Analysis, and Dissemination of Results for Educational Research.* Unpublished manuscript, University of Arkansas, Fayetteville, AK.

National Science Foundation (1999). *Dual announcement for systemic nitiatives research studies and Rural Systemic Initiatives evaluative studies* (NSF 99-95). Arlington, VA.

Westat*McKenzie Consortium. (1998). *The National Science Foundation's Statewide Systemic Initiatives (SSI) Program: Models of reform of K-12 science and mathematics education.* Washington, DC.

*Webb, N. (1998). *Study of the impact of statewide systemic initiatives.* Unpublished manuscript, University of Wisconsin—Madison, WI.

Williams, L. S. (1996, October). *The systemic initiatives: Salient Issues.* Presentation at the Fall Principal Investigator Project Director Meeting, Washington, DC.

*Yin, R. (1999). *Studying statewide systemic reform.* Unpulished manuscript, COSMOS Corporation, Bethesda, MD.

Yin, R. (2001). *Cross-site evaluation of the Urban Systemic Program—First annual report.* COSMOS Corporation. Bethesda, MD.

Zucker, A., Shields, P., Adelman, N., Corcoran, T., and Goertz, M. (1998). *A report on the evaluation of the National Science Foundation's Statewide Systemic Initiatives (SSI) Program.* (NSF 98-147). National Science Foundation. Arlington, VA.

CHAPTER 4

Musings on Science Program Evaluation in an Era of Educational Accountability

Dennis W. Cheek

INTRODUCTION

Science education programs are always situated in the much larger, loosely coupled, education system comprised of national, state, and local dimensions. Science education programs in any given locale vary substantially due to the confluence of many contextual factors, including the nature and extent of educational accountability systems. Evaluation of school science programs has to be undertaken with these broader systems in mind. Explicit recognition of these dimensions and studies of their intended and unintended effects can contribute substantially to improving current thinking and practice about both educational accountability and school improvement. This chapter provides an overview of contemporary educational

Dennis Cheek, Office of Research, High School Reform and Adult Education, Rhode Island Department of Education, University of Rhode Island, 255 Westminster Street, Providence, RI 02903-3414.

Evaluation of Science and Technology Education at the Dawn of a New Millennium, edited by James W. Altschuld and David D. Kumar, Kluwer Academic / Plenum Publishers, New York, 2002.

accountability in American K-12 education, discusses the limitations of current conceptions and models, portends changes for the future in the accountability arena, and suggests issues that evaluators may wish to address as they formulate and implement evaluation plans.

THE RISE OF NEW AMERICAN ACCOUNTABILITY SYSTEMS

Public education in the United States has been plagued by perceptions of its inadequacies since its earliest days (Cremin, 1990; Krug 1969, 1972; Public Agenda, 2001). The latest panacea to the perceived problems in American K-12 education is expanded educational accountability. This movement is not solely an American phenomenon as many nations and international organizations have created or refined such systems in recent years (e.g., OECD 2000; Black & Atkin, 1996; Hurd, 1997; Schmidt *et al.*, 1997a, 1997b; Osborne & Pilot, 2000; Chen, 1999).

Accountability has typically been approached as an outcomes or educational output oriented exercise. This usually involves the promulgation of standards of one form or another as a starting point for system redesign (Wolf, 1998; Bybee, 1998). These standards attempt to specify what all students should know and be able do to at particular points in their journey through K-12 schools. The standards are predictably followed by annual measurement of progress toward their attainment employing student cohorts, state and/or national testing instruments, and frequently, individual student scores tied to important consequences, e.g., granting of graduation diplomas, advancement to the next grade, access to course enrollment, participation in summer school (Finn, 1991; Coleman *et al.*, 1997; Research and Policy Committee, 2001).

This idea of testing to sort and "reward" students in a consequential manner is an ancient idea that appears to have its roots in the civil service examinations of Imperial China. Exam results alone determined which males throughout China were accorded the enviable position of becoming a life-long civil servant with its many attendant benefits. Historical evidence shows that many of the unintended consequences of contemporary high stakes testing occurred millennia ago in China including rampant cheating, nonuniform preparation programs and testing situations that favored some candidates over others (Mote, 1999; Elman, 2000). These Chinese practices around high-stakes examinations may have been brought to the West during the British colonial era. They certainly have been enshrined in the British Commonwealth schools for well over a century and a half (Roach, 1971).

American policy makers have given little attention to the dark underside of high stakes testing. Formidable consequences have been documented in contemporary Japan, Korea, and Taiwan (Zeng, 1998). These include massive after-school programs that teach to the test, extensive and severely restrictive educational sorting systems based solely on exam results, and documented instances of student suicides following the annual release of exam scores.

The accountability design path followed in American schools is often exactly backward to the ideal model with a current test either driving the creation of standards or proceeding independent of new content standards. This severe misalignment sends highly confusing messages to educational practitioners (Webb, 1999). Better alignment of the tests to standards, however, would not guarantee improved results. There is clear evidence that science curriculum materials used in American schools are incongruent with science content standards promulgated by the National Research Council and AAAS's Project 2061 (National Research Council, 1996; Schmidt et al., 1997a, 1997b; American Association for the Advancement of Science 1993). Project 2061 of AAAS has recently conducted an in-depth content analysis of current middle school science textbooks and high school biology textbooks using the *Benchmarks for Science Literacy* as the content standards (American Association for the Advancement of Science, 1993, 1999, 2000). They found all of the middle school science textbooks and the overwhelming majority of the high school biology textbooks seriously deficient. Key highlights include errors of content, cursory treatment of important concepts and topics, and materials that provide students with fleeting glimpses of the true nature of science and fundamental science concepts. Quality curriculum materials that focus on conceptual understanding are a necessary component of a constructivist approach to student science learning (Mintzes et al., 1999).

Changes in tests and curricula are only part of the accountability puzzle. The ascendancy of the accountability agenda brought considerable changes to federal education law including the reauthorized Improving America's Schools Act (IASA, formerly titled the Elementary and Secondary Education Act-ESEA), laws that govern special education (IDEA—Individuals with Disabilities Education Act), and career and vocational education (Perkins III). Enshrining this new accountability system in legislation took Congress over eight years and involved intense lobbying by large American corporations and special interest groups (Jennings, 1998). There were assumed linkages among the American economy, a "failing" educational system, and declining economic competitiveness in the global marketplace (e.g., National Commission on Excellence in Education, 1983). The U.S. economy seems to have recovered quite well over the past twenty years despite little evidence of any dramatic improvements in K-12 education—suggestive of the complex nature of this relationship (Levin,

1998; Cheek, in press). The resulting legislation is daunting in size, complex in nature, difficult to interpret, and contradictory from one program to another—a logical result of a political process where many different interests have to be heard and represented in some manner in the final product. The U.S. Department of Education has subsequently issued voluminous additional regulations that "interpret" and operationalize the legislation. A considerable number of assumptions appear to underlie the legislation and its accompanying regulations including:

1. Defining clear content standards for teachers and performance standards for students will result in a higher quality education for all students
2. There is a straightforward relationship between teaching and learning
3. There is a straightforward relationship between learning and testing
4. Publicly announcing results will spur continuous improvement within American schools
5. Considerable gaps in student achievement when measured against absolute standards and among student subgroups will disappear entirely and relatively homogeneous knowledge and skills are desirable
6. Jobs with sufficient conceptual challenge and ample further educational opportunities will exist in the workplace for these enormously expanded numbers of higher caliber K-12 graduates
7. There will be no adverse effects from the implementation of expanded educational accountability

This increased accountability for federally funded programs administered by states have fueled high-stakes student testing, public reporting, and sanctions and rewards. The urgency around this issue has been accelerated by the sheer amount of dollars in state budgets that are devoted to K-12 education. (The relative percentages of education versus non-education expenditures have fluctuated up and down considerably from one state to another depending on the local economy.) In the small state of Rhode Island, for example, expenditures for K-12 education from combined state, federal, and local resources is a $1.7 billion annual investment.

Forty-five of the 50 states in America now publish school and/or district reports on an annual basis—when only a handful conducted such activities two decades ago (see www.ccsso.org for links to all fifty state education department websites). Twenty-seven states publicly rate performance or at least identify low performing schools. Eighteen states have been given direct powers to intervene in local schools who are viewed as chronically underperforming and 20 states provide some form of reward to schools who perform in a manner judged acceptable. (Education Week, 1999, 2000, 2001).

State accountability systems for education are designed to address the twin perceptions of chronically low student performance on state administered tests and the perceived failure of schools to adequately prepare students for life, living and further education in the twenty-first century. These policy formulations coupled with technical functions of collecting, analyzing and publishing data along with a continuum of state intervention procedures are intended to bring about desired changes in schools and populations of students.

Science program evaluation in this new milieu of high stakes testing and accountability becomes quite important. Evaluators can help individual schools better understand how their current actions and programs might or might not affect student performance on externally administered exams and other indicators (e.g., McNeil, 2000; cf. Skrla, 2001). Savvy evaluators can marshal quantitative and qualitative measures over time that paint rich portraits of how student learning can be improved by specific teaching strategies and the use of particular program materials. Hopefully, the evaluation community can also insert itself into a more prominent role in helping state and federal policy makers to design accountability systems that are fair, reliable, valid, and important in terms of what they measure (e.g., Miller, 2001; Horn *et al.*, 2000).

The charges that have been levied against education, the "new" reforms initiated, and the "results" achieved to date are not really "new" at all (Tyack & Cuban, 1995). America has gone through many cycles of educational reform, each of them never achieving much of what their creators intended and all of them leading to yet new cycles of reform rhetoric and action. Understanding the reasons why this is so can be aided by thinking about these accountability systems as technologies of social control (cf. Madaus, 1990).

ACCOUNTABILITY SYSTEMS AS TECHNOLOGIES OF SOCIAL CONTROL

Kline (1985) of Stanford University suggests that "technology" embraces a vast set of concepts, artifacts, and systems designed by humans to meet human needs and achieve desired purposes. Technology can be discussed in four major ways: 1) as artifacts or hardware, e.g., pencils, microscopes, antiballistic missiles, a test; 2) as sociotechnical systems of production, e.g., an automobile assembly line or elementary and secondary education where some end product(s) is produced; 3) as technique or methodology, e.g., the skills, knowledge, and general know-how to rebuild an engine, engage in oil painting or solve a math problem; and 4) as sociotechnical systems of use that are linked to a specific artifact(s), e.g., an airplane presupposes a much wider system of rules and regulations, licenses and trained pilots, passengers and/or cargo, maintenance, airports, manufacturing facilities, and air traffic control. In a similar manner, diplomas conjure up images of a system of teaching, learning some body of knowledge and honing a set of skills, authentication of that learning by some officially recognized group authorized to sanction that it has occurred, and a system of rewards that follow receipt of such documents.

Kline's ways of discussing technology remind us that state accountability systems involve a complex set of artifacts. These artifacts include criterion and norm-referenced tests, demographic data, and school data about measures such as dropouts, average daily attendance, teacher grievances, and disciplinary actions. Such artifacts are increasingly melded together into published reports (some states referring to them as "report cards") from which policy makers and others are encouraged to make value judgements about school and district "quality." Sometimes a state goes even further and specifically labels particular schools on an evaluation continuum from "excellent" or similar commendations to "failing" or terms with similar negative connotations. In Ohio for example, schools are rated as effective, continuous improvement, academic watch, and academic emergency.

State policy makers have created these formal evaluation systems often without considering the value assumptions inherent in their design. Some key underlying assumptions about state accountability efforts as presently constituted might include:

(1)   educational quality can be determined from a fixed (and generally small) number of measures, collected within one or more brief time series

(2)   students perform to the best of their abilities on state administered tests

(3)   tests are well connected to local school curriculum sequences and substantially correlate with individual student grades (i.e., other evidence of student achievement collected independent of the state assessment system)

(4)   publishing information on the measures the state selects, in and of itself, will serve as a major stimulant to school improvement

(5) schools will routinely supply *accurate* and *timely* information to the state upon request

(6) school-based strategic planning, informed by the kinds of measures the state collects and publishes, will lead to better student performance on state tests and improvements in other measures valued by the state

(7) measures that the state values are more important to the individual school than any competing set of measures which might lead the school to alternative courses of action

(8) there is common and pervasive understanding of the state's educational reform agenda from the superintendent's office and school committee down to the level of the individual classroom teacher and the agenda is sufficiently detailed to suitably guide local planning and action

Many more assumptions can undoubtedly be generated. This list is sufficient to make the point that several of these assumptions are not supported by research findings about school reform to date in the United States or elsewhere (cf. Broadfoot, 1996; Fink, 2000; Mungazi, 1999; National Research Council, 1999; Sarason, 1996; Vinovskis, 1999). Analyses of school curricula, for example, have shown a frequent mismatch in both content and sequence with state assessments. Many state assessments are poorly aligned with state promulgated content standards. How much effort students, especially older ones, put forth on state tests, is still a subject of debate. Teacher anecdotes indicate that the tests are not taken seriously by large numbers of students. Audits of school-reported data to date have generally found significant reporting errors (Mayer *et al.*, 2000).

Despite these documented shortcomings, state government will continue to act in the interests of both their younger and older citizens to constantly improve the educational experiences and outcomes of public education. Expenditures of public monies on a large scale will always prompt attendant interests by federal, state, and local government in the impact of their investments. School reform, therefore, will remain a permanent fixture of the educational landscape. Recognizing accountability systems as technologies of social control, however, holds promise of informing our decision-making processes and perhaps mitigating harm.

Some observers and participants are beginning to question whether we have lost sight of our goals in these expansive accountability systems. The current Washington state superintendent who pushed for the adoption of a rigorous state assessment system has already had some second thoughts, stating "I now seriously question where we are going with a driver that seems drunk with power." (Billings, 2000: 2) The American Education

Research Association (AERA) has viewed the recent wave of increased state and federal accountability systems based heavily on individual student test results with alarm and issued a formal policy statement regarding high-stakes testing in July of 2000 (AERA, 2000). School district officials, building administrators, and teachers still evince a lack of understanding about the nature and meaning of these new educational standards despite intensive efforts in their promulgation and much written guidance regarding how to approach their implementation (e.g., Donmoyer, 1998; Elmore & Rothman, 1999; Spillane & Callahan, 2000).

Evaluators of science programs at all levels need to build within their evaluations means to measure and evaluate the role of these accountability mechanisms in promoting or inhibiting good science teaching and learning. Policy makers need quality evidence of the impact of their decisions in classrooms and in students' lives that can never be fully or adequately captured by existing state accountability systems. Much better evidence also needs to be collected about the actual processes by which teachers come to thoroughly understand content standards and the processes required to translate knowledge of these standards into effective instructional and assessment practices with their students. We are a long way from understanding these matters with any clarity.

NATIONAL AND INTERNATIONAL TESTING

Even as states and many larger school districts have created massive machinery for educational accountability, the federal government has increased the pace and breadth of its assessment efforts to compare states to one another and the United States to other nations. Since 1969, the National Assessment of Educational Progress (NAEP) has conducted periodic assessments of representative samples of students in participating states in reading, mathematics, science, writing, history, geography, and other fields. The Bush administration proposed in 2001 that NAEP be administered every year in reading, writing and mathematics and on a more consistent basis in other content areas. NAEP has become one standard by which achievement in different states is compared and is presently the best instrument in use for this purpose (despite its many limitations). Interestingly, virtually all of the 15 innovative features recommended for NAEP during its design phase in the sixties have been minimized, violated or entirely dismissed over the past three decades. For example, one recommendation was that probability sampling from private as well as public schools be undertaken as well as students who were not in school at the target age (e.g., migrant children, school dropouts).

Additionally, efforts by panels of judges to define the various achievement levels (advanced, proficient, or basic) in an adequate and defensible manner have proven difficult (Jones, 1996). Many conclusions frequently drawn from NAEP results, therefore, cannot withstand close scrutiny.

While NAEP is still the best national testing instrument, it has more than its share of problems. Attempts to link its results with those of other forms of standardized tests—both criterion and norm-reference based— have proven psychometrically elusive (Feuer et al., 1999). However, NAEP scores have routinely been cited over the past two decades in support of expanding national and state accountability systems.

The Third International Mathematics and Science Study (TIMSS) sponsored by the International Association for the Evaluation of Educational Achievement and principally supported by the United States, has also fueled the drive toward increasing accountability. TIMSS is the largest comparative study of educational achievement ever undertaken (see Schmidt, this volume). Tests in mathematics and science were administered to selected samples of students in more than 40 nations involving more than half a million students and over 30 languages. The U.S. Department of Education has released the TIMSS reports in regular press conferences and highlighted them in virtually every national meeting in which it has participated in the past four years (e.g., Beaton et al., 1996; Martin et al., 1997; Mullis et al., 1998; National Center for Education Statistics, 1998). It also funded a host of ancillary products to help states, districts, and schools use these materials and administered a set of follow-up TIMSS-R ("Repeat") assessments in selected nations and states with results released in fall of 2000 (OERI, 1997). The meaning that is to be attached to TIMSS results for U.S. policy makers has been the subject of considerable and ferocious scholarly debate. Many key issues still appear unresolved or unresolvable.

TIMSS results have little import for school-level practitioners despite the federal government's attempts to make them think otherwise (Baker, 1997; Bempechat & Drago-Severson, 1999; Bracey, 1997, 2000a, 2000b; Forgione, 1998; Stedman, 1997a, 1997b). The reasons for this are varied but include the lack of alignment between TIMSS items and U.S. curricula in mathematics and science, the use of national samples that contribute nothing to understanding local or state performance, and the long delay in release of reports. TIMSS results have been highly influential in debates about federal funding for K-12 science and mathematics education and are frequently invoked in state deliberations about science and mathematics funding.

Evaluators can play a role in these debates and the overly broad use of TIMSS results in policy discussions of education reform by explaining the purposes and limitations of such assessments to policy makers at national, state, and local levels. Using selected released TIMSS items, coupled with other types of science assessment items, evaluators can help all parties to better understand what individual items reveal or fail to reveal about student learning and what batteries of items reveal or fail to reveal about students, teachers, and school systems.

Current Design Flaws

Much is known today about the various tradeoffs embodied within particular types of test items and types of tests. The prohibitive costs associated with extensive testing using a wide ranging set of measures and assessment & across an expansive set of content and process domains means we will never witness tests that fully comply with the best current thinking about assessment. Contemporary tests, like all tests that have come before them, are measurement tools that represent compromises among sound principles of testing, fiscal constraints, development constraints, and reporting constraints. Understanding the nature of some of these issues and their interplay is helpful to evaluators and policy makers.

At the state level, tests have become so pervasive that NAEP had great difficulty in recent years getting states and/or selected school districts and schools to participate sufficiently to provide a valid sample. Fewer than 40 states participated in the last NAEP testing in reading and interviews with many commissioners of education and local officials cited testing burden due to increased state accountability systems and lack of usefulness of the results as the principal reasons for lack of interest. The National Center for Education Statistics released an RFP for a contractor to work on increasing NAEP participation in the future by helping NAEP and the National Assessment Governing Board to develop better strategies and incentives to induce participation. The impact of the Bush administration's proposal to increase NAEP testing in frequency and breadth on an annual basis must be watched with interest.

State testing systems vary widely in scope, validity, reliability, and comparability but still serve some useful purposes (Blank & Langesen, 1999; Education Week, 1999, 2000, 2001; Ansley, 1997; Olson, 2001). Standards developed with considerable care and debate by testing and program evaluation experts over many years are frequently not realized in the designs and implementation of state accountability and testing systems (AERA *et al.*, 1999, Joint Committee on Standards for Educational Evaluation, 1994, Educational Testing Service, 2000). Experts on accountability continue to hope that someday, somewhere, a district or state accountability system may arise that will possess sufficient complexity and technical rigor that the results can assuredly be trusted (Baker & Linn, 2000). Success at this venture would undoubtedly require substantially higher levels of investment in state testing and given revenue sources at state and national levels and competing needs in education, these increases will likely never be realized. The field is also heavily influenced by arguments about face validity. Despite their high development and administration costs and nontrivial psychometric problems, an increasing number of tests feature performance-based tasks in whole or in part (Fitzpatrick & Yen, 1999; Lawrence *et al.*, 2000; Parkes, 2000; Webb *et al.*, 2000). Practitioners and many policy makers favor them because they much more closely resemble what teachers do or "should do" in science classrooms (Miller, 1999, Loadman & Thomas, 2000). The argument has been made that these tests are "worth teaching to" and that having a wealth of performance-based tasks on state tests will "drive" teachers to adopt performance-based teaching and performance-based local assessment practices (e.g., Tucker & Codding, 1998).

A more central design flaw to the current plethora of content and performance standards and tests is the inappropriate cognitive demands that the standards and test items sometimes make upon students from a set of wide development ranges (Metz, 1997). Partly this is a problem of educational research, as the following illustration from science education research demonstrates. Project 2061 made a serious attempt to ground its *Benchmarks for Science Literacy* in the science education research literature with a special chapter about children's conceptions about the natural world. For most of the benchmarks there was no relevant research and it was quite fragmentary in areas where there had been some work (AAAS, 1993). The intervening years have seen an increase in research of this kind. The gaps are still quite wide, however, between what the two leading science standards documents espouse that all children should know and be able to do and documented evidence that such pronouncements are developmentally appropriate and attainable (Roseman, 2001). This is an extremely fertile ground for the work of science education evaluators.

The science content standards currently promulgated, therefore, have only a modest research base underlying them. In addition to the problem of developmental appropriateness, we can also note that none were derived from empirical investigations of the kinds of scientific literacy people exhibit or seem to require in everyday life and work. Science standards, as well as those in other fields, were developed by teams of "experts" who make their living in this arena. These experts in turn, then decide what everyone else who is not steeped in this arena needs to know in order to be considered "literate." This standardization of knowledge can be legitimately questioned on a number of fronts (e.g., Vinson *et al.*, 2001). Additionally, in the rush to get all students to meet the standard, we may be further reinforcing to students that science is that which is passively received and mastered rather than a method of acquiring knowledge that has its roots in skepticism (Franknoi, 1999). Issues like these need to be further explored by science program evaluators as they frame evaluations at various levels of the system.

Few have given enough thought to what explicit model(s) of learning might be required to best position a student for lifelong science learning rather than the learning of the moment (Linn & Muilenburg, 1996). On the one hand, it is clear that a substantial number of younger children can engage in more demanding cognitive tasks than many teachers and policy makers believe (Metz, 1995). On the other hand, the link between cognition and learning, despite many recent advances, is still far from clear in terms of what exactly a teacher can or should do to best position an individual student for learning (Bransford et al., 1999; Kuhn, 1997, 1999; Cobb & Bowers, 1999).

Coupled with these developmental and cognitive dilemmas is the clearly differentiated quality in school instruction, curriculum, and facilities experienced by students (Firestone et al., 2000; Lee, 1999). The standards movement in the view of liberals was intended to be a means by which greater uniformity of learning environments could be achieved as we focused on ALL students having access to a quality curriculum. Conservatives generally have supported standards with arguments having to do with accountability for results, the high costs associated with education, economic competitiveness, and perceptions about a lack of school quality generally as compared to former decades (cf. the perceptive comments of contemporary Dewey disciples Vinson et al., 2001).

Insufficient opportunity to learn can be written in bold letters across virtually all "failing" schools and the real question is: "Who has failed?" Student failure in these schools is often an indicator of the explicit failure of the educational system to provide safe and adequate learning facilities, qualified teachers, abundant and current learning materials, and adequate support services. Evaluators must attend to these contextual facets of science education both in the conduct of their studies and in interpreting their results for various parties.

CHANGE AT THE LOCAL LEVEL

The arena where all of these accountability systems bear down most heavily is within a single school, and in particular, a classroom. It is reasonably clear from NAEP and other student and teacher surveys that science curricula in schools and teacher practices do not resemble what is advocated in national science reform documents (Blank & Langesen, 1999; Council of Chief State Schools Officers, 2000). The National Science Foundation, in recognition of this fact, has recently started a series directed to district and school-level practitioners regarding science inquiry and quality science programs (Division of Elementary, Secondary, and Informal Education, 1997, 1999). Certainly the massive NSF investment in recent years in Statewide Systemic Initiatives, Urban Systemic Initiatives, Rural Systemic Initiatives, and Local Systemic Initiatives was partly in response to the fact that reform was sporadic, unpredictable, and quite deep in only a few places. Evaluations to date of these efforts suggest that while some of these investments have made an impact, it has been much more modest than had been anticipated (Knapp, 1997; Blank, 2000). The grand vision of using fairly limited federal dollars as levers for massive structural changes in educational systems (systemic reform) has proven to be largely a pipe dream. State and local politics, varied expectations, divergent motivation, competing goals, and limited additional resources have exerted their combined influences to minimize the impact of reforms.

The educational landscape is busy with the development of yet more wonderful science curricula and expanded professional development opportunities for teachers (e.g., Black & Myron, 1996, North Central Regional Educational Laboratory, 2000). One wonders whether we might do better to declare a nationwide moratorium on the development of new science materials. Funds normally expended in these activities could be reprogrammed into implementation strategies for existing good curricula and institutionalization of quality professional development in mathematics and science. NSF has taken a preliminary step down this road with their funding of implementation centers for mathematics and science materials produced in part with NSF funds.

Science education is still plagued by a lack of focus. Many different meanings regarding "science education" and its purposes can be found implicitly or explicitly within the materials themselves and the system by which they are delivered to students. Roberts (1998) has identified seven distinct curriculum emphases in science education materials:

- everyday coping (applied science in everyday life)
- structure of science (science as an intellectual enterprise)
- self as explainer (student ownership of the process of explaining science)
- scientific skills development (reliable scientific knowledge is acquired through application of proper methods)
- solid foundation (learning science content now as the foundation for further study of science)

- correct explanations (how scientists understand the world is correct and to be accepted)
- science, technology and decisions (scientific knowledge must be interwoven with other forms of human knowing to make appropriate decisions on personal and societal levels)

These particular emphases can be seen in contemporary science materials and all materials emphasize some of these orientations to the exclusion of others. The loosely coupled American educational system itself also contributes to the contemporary semantic confusion regarding science education and education generally (Tye, 2000).

BETTER ASSESSMENT OF STUDENT ACHIEVEMENT

Science assessment and program evaluation in science have major roles to play as we move from the current situation toward a more desirable one. Teachers in this era of educational accountability seem more interested than ever in matters related to determining whether a student has truly mastered a concept or principle. Publications targeted to teachers more frequently address this issue (e.g., Cole, 2000; Stine, 2000). The National Science Teachers Association has developed several resources in recent years and convention sessions and workshops for science teachers at all levels of the system often have assessment issues as their focus (e.g., Doran *et al.*, 1998). Improving the teacher's use of good assessment strategies in a mixed method approach that is driven by educational purpose is clearly needed. But additional tools must be developed in concert with assessment and evaluation experts if greater progress is to be made.

Practices such as baseline assessment and value-added approaches to individual student achievement need to become as common in America as they are becoming in other nations such as Great Britain (cf. Sanders *et al.*, 1997; Lindsay & Desforges, 1998; Tymms, 1999). These approaches involve collecting information to generate profiles of students' existing knowledge and ability and then predicting their future performance on commonly employed assessment instruments, including accounting for possible errors in the prediction model. These value-added systems enable teachers and administrators to see how students actually perform when you take into account their prior knowledge and achievement (always the most powerful predictor of future performance, accounting for about 50% of the variance). The impact of individual teachers on cohorts of students can be effectively seen in these models. Multiple years of data place evaluation of individual teacher performance on a much firmer footing than systems currently employed, often resulting in teachers voluntarily seeking specific ways to improve their teaching as they become convinced that the trends seen in these ever-growing data sets are real.

A student's individual progress against a whole variety of standards needs to be monitored and tracked in a regular manner as many computerized instructional management systems now permit (Masters *et al.*, 1999; Schacter *et al.*, 1999; Wilson & Sloane, 2000). Adaptive, computerized testing systems modeled after the immensely successful ASVAB of the American Armed Forces need to become ubiquitous in school settings (Sands, Waters, & McBride, 1997; Bennett, 1998). Until systems like these become widespread, much of the rhetoric about performing to standards is just that. If we truly believe that the standards we have set are important for all students to achieve, then we would give students as many opportunities as needed to demonstrate that they have mastered a particular standard. Instead we assess them at usually a single point in time and then push them to attempt attainment of a further set of standards, whether they mastered the prior standards or not. Adaptive computerized testing would enable students' performance against standards to be tracked, regularly tested using a wide array of items and performance levels, and monitored and portrayed in an effective and efficient manner.

Testing itself will continue to evolve in the twin arenas of classical test theory and item response theory (IRT). The former focuses on deriving a total score from an assessment battery and utilizing such a score in an informed manner. IRT provides a fine-grained analysis on the contribution of classical theory by focusing on how an individual performs on a single item or question and its relationship to the trait being measured by the test as a whole. Until recently, use of IRT was greatly limited by issues around the computations themselves and computer programs with sufficient power and sophisticated to perform the calculations. Computerized tools increasingly enable practitioners to routinely utilize techniques like latent trait measures, item response theory, Rasch measurement theory, partial credit models, rating scale analysis, bias measurement, equating, and item analysis to improve our understandings of what is being measured and the fidelity of the measurement to the desired outcome (Masters & Keeves, 1999).

IMPROVING EVALUATION OF SCIENCE TEACHING AND LEARNING

Ongoing evaluation of school science programs must move considerably beyond where it is today. One of the first problems to be surmounted is helping state and local policy makers, teacher union leaders, rank and file teachers and administrators, and others to see the need for such evaluation efforts. It is an interesting historical note that the American Educational Research Association had its origins in the National Education Association (Grant & Graue, 1999). Yet you would never guess this historical connection when you visit either an AERA convention or a NEA convention since the former hardly ever includes practitioners in its membership ranks or sessions and the latter hardly ever includes researchers or research in it deliberations. AERA is presently considering a substantial reorganization of its national headquarters in order to bring AERA expertise more fully into national and state policy-making arenas. AERA collaborated with the American Psychological Association and the National Council for Measurement in Education to produce a set of standards for evaluation in 1994.

Some federal agencies recognize the important role that evaluation can play in better informing policy and practice in K-12 education. NSF produced one of the first evaluation guides targeted to non-evaluators that explicitly addressed science, mathematics, engineering, and technology education (Frechtling, 1993). The expansion of state accountability systems has also spawned a spate of books targeted to district officials and school improvement teams dealing with evaluation and assessment issues (e.g., Leithwood & Aitken, 1995; Levesque *et al.*, 1998; Wahlstrom, 1999). They show practitioners how to think about data and suggest ways in which data can be manipulated to answer questions of concern to practitioners. While they lack the sophistication of recent textbooks in the field (e.g., Mohr, 1995; Shadish *et al.*, 1991) they are a major move forward and can serve as very useful professional development tools for evaluators to use in their work with schools and school districts.

At the local level, evaluators must earn practitioners' respect and trust by addressing issues that are viewed as highly relevant. This might require the evaluator to work with teams of teachers in action research projects to determine whether a particular program is worthwhile to adopt school-wide or district-wide. Action research also can be effectively undertaken in conjunction with district supervisors to build a strong case for professional development programs and to better plan for those efforts using locally derived data. Teachers, principals, and other school-based staff need much deeper understanding of basic principles of assessment and evaluation as they apply to classroom settings and individual students. Administrators in particular need to become much more familiar with national guidelines about assessment and evaluation practices.

Increasingly schools are instituting formal mentoring and induction programs. Evaluators can help schools and districts understand how their program is currently functioning, what kinds of changes might be worthwhile, and the impact of those adjustments. These studies can be grounded in a series of national studies of mentoring and induction commissioned by the Education Commission of the States with funding from the U.S. Department of Education.

Schools need help in the complex process of understanding and effectively using the various standards documents that have been promulgated. They require relevant data to help guide their determinations not only about what to teach but what to ignore in both standards and in written materials. Most schools struggle with issues around the alignment of curriculum, instruction, and assessment. Evaluators can play an effective role in helping practitioners work through the relevant issues in a satisfactory manner. A further pressing need in the curriculum is to assist schools to develop integrated mathematics, science, and technology education and better connect science education with the social sciences and other disciplines. This work involves the interplay and overlap of content standards such as those promulgated by the National Council of Teachers of Mathematics (2000), National Research Council (1996), American Association for the Advancement of Science (1993), and International Technology Education Association (2000). More importantly, it requires translating some of these national connections into local analogues since most teachers and administrators are focused principally on their own state's content and performance standards and not the national ones. The pervasive influence of educational technology and its potential across the science curriculum to improve teaching and learning is another fertile area in which to assist schools. The systematic study of various implementation strategies and their impact on student learning of concepts about computers and related devices *and* science content knowledge—both of which are essential in the technological society we inhabit—are critical components of quality science programming.

Interestingly, one key target audience for all these national and state tests results, parents, appear to be leagues behind policy makers and school personnel in their knowledge and understanding of test results (Shepard *et al.*, 1995). Evaluators can play an important role in helping give voice to the needs of parents and community members in better understanding these issues. They can also routinely build evaluation of the views of parents and community members into their evaluation designs.

The entire arena of science education research also needs to be further enhanced by the systematic work of evaluators. One area where this is especially important is the use of experimental and quasi-experimental methods on a large scale to determine the effectiveness of particular programs and practices in science teaching and learning. The Campbell Collaboration (www.gse.upenn.edu) is an international organization recently created to advance the science of systematic reviews and their use in education, social welfare, and criminal justice worldwide. Randomized controlled trials and quasi-experimental methods conducted in a widespread and systematic manner help minimize bias in research (Boruch *et al.*, 2000; Bickman, 2000). When coupled with contextual data, systematic reviews present very powerful evidence for the effectiveness of various practices that have been advanced in science education literature.

As increasingly sophisticated district student information systems develop, evaluators need to work closely with producers and users of such systems to ensure the worthiness and adequacy of the automated tools for analysis available within these instructional management systems. Expert systems will likely come more fully into play in education as they have in other arenas such as medicine and engineering.

CONCLUSION

Evaluators of science programs in school settings need to be aware of the much larger accountability context within which school science is situated. Any evaluations of science programs in schools must be situated in a manner that speaks to the larger accountability agenda of the district and the state. Evaluators need to be seen as providing value to the educational system in terms of the skill sets, tools and techniques, and understandings they bring to school science. School personnel and policy makers, working in concert with experienced evaluators, can derive many useful insights into science teaching practices and student learning that go well beyond insights gained from traditional accountability measures. Evaluators also have a social responsibility to help inform the use and application of accountability measures as they relate to science education.

REFERENCES

American Association for the Advancement of Science (1993). *Benchmarks for Science Literacy.* Oxford University Press, New York.
American Association for the Advancement of Science (September 28, 1999). Heavy books light on learning: Not one middle grades science text rated satisfactory by AAAS's Project 2061. *AAAS Project 2061 Press Release*, Washington, DC.
American Association for the Advancement of Science (June 27, 2000). Big biology books fail to convey big ideas, reports AAAA's Project 2061. *AAAS Project 2061 Press Release.* Washington, DC.
American Educational Research Association, American Psychological Association, and National Council on Measurement in Education (1999). *Standards for Educational and Psychological Testing.* AERA, Washington, DC.

American Educational Research Association (2000). AERA position statement concerning high-stakes testing in PreK-12 education with comments by five researchers. *Educational Researcher*, 29(8):24–29.

Ansley, T. (1997). The role of standardized achievement tests in grades K-12. In Phye, G. D. (Ed.), *Handbook of Classroom Assessment: Learning, Achievement, and Adjustment*. Academic Press, San Diego, pp. 265–285.

Baker, D. P. (1997). Response: Good news, bad news, and international comparisons: Comment on Bracey. *Educational Researcher*, 26(3):16–17.

Baker, E. L., and Linn, R. L. (Spring 2000). A field of dreams. *The CRESST Line: Newsletter of the National Center for Research on Evaluation, Standards, and Student Testing*, 1–3. Center for the Study of Evaluation, UCLA, Los Angeles, CA.

Beaton, A. E., Martin, M. O., Mullis, I., Gonzelez, E. J., Smith, T. A., and Kelly, D. L. (1996). *Science Achievement in the Middle School Years*. Lynch School of Education, Boston College, Chestnut Hill, MA.

Bempechat, J., and Drago-Severson, E. (1999). Cross-national differences in academic achievement: Beyond etic conceptions of children's understanding. *Review of Educational Research*, 69(3):287–314.

Bennett, R. E. (1998). *Reinventing Assessment: Speculations on the Future of Large-Scale Educational Testing*. Policy Information Center, Educational Testing Service, Princeton, NJ.

Bickman, L. (Ed.) (2000). *Validity & Social Experimentation. Donald Campbell's Legacy, Volume 1*. Sage Publications, Thousand Oaks, CA.

Billings, J. A. (2000). Guest editorial. Two words are inextricably connected. *Technos Quarterly for Education & Technology*, 9(2):2.

Black, P., and Atkin, J. M. (Eds.) (1996). *Changing the Subject: Innovations in Science, Mathematics, and Technology Education*. Routledge, New York.

Blank, R., and Langesen, D. (1999). *State Indicators of Science and Mathematics Education 1999*. Council of Chief State School Officers, Washington, DC.

Blank, R. (2000). *Summary of Findings from SSI and Recommendations for NSF's Role with States: How NSF Can Encourage State Leadership in Improvement of Science and Mathematics Education*. Council of Chief State School Officers, Washington, DC.

Boruch, R., Snyder, B., and DeMoya, D. (2000). The importance of randomized field trials. *Crime and Delinquency*, 46(2):156–180.

Bracey, G. W. (1997). Rejoinder: On comparing the incomparable: A response to Baker and Stedman. *Educational Researcher*, 26(3):19–25.

Bracey, G. W. (2000a). *Bail Me Out! Handling Difficult Data and Tough Questions about Public Schools*. Corwin Press, Thousand Oaks, CA.

Bracey, G. W. (2000b). The TIMSS "final year" study and report: A critique. *Educational Researcher*, 29(4):4–10.

Bransford, J., Brown, A., and Cocking, R. (Eds.) (1999). *How People Learn: Brain, Mind, Experience, and School*. National Academy Press, Washington, DC.

Broadfoot, P. M. (1996). *Education, Assessment and Society*. Open University Press, Philadelphia, PA.

Bybee, R. W. (1998). National standards, deliberation, and design: The dynamics of developing meaning in science curriculum. In Roberts, D. A. and Ostman, L. (Eds.), *Problems of Meaning in Science Curriculum*. Teachers College Press, New York, pp. 150–165.

Cheek, D. W. (in press). Education and economic growth. In Smelser, N. J. and Baltes, P. B. (Eds.), *International Encyclopedia of the Social & Behavioral Sciences*. Pergamon Press, New York, 26 volumes.

Chen, D. (1999). Confronting complexity: A new meaning to world literacy. In Chaisson, E. J. and Kim, T.-C. (Eds.), *The Thirteenth Labor: Improving Science Education*, pp. 87–100. Gordon and Breach Publishers, Newark, NJ.

Cobb, P., and Bowers, J. (1999). Cognitive and situated learning perspectives in theory and practice. *Educational Researcher*, 28(2):4–15.

Cole, N. S. (2000). Determining what is to be taught: The role of assessment. *ENC Focus: Assessment that Informs Practice*, 7(2):34.

Coleman, J. S., Schneider, B., Plank, S., Schiller, K. S., Shouse, R., Wang, H., and Lee, S.-A. (1997). *Redesigning American Education*. Westview Press, Boulder, CO.

Council of Chief State School Officers (2000). *Using Data on Enacted Curriculum in Mathematics & Science. Sample Results from a Study of Classroom Practices and Subject Content*. Washington, DC.

Cremin, L. A. (1990). *Popular Education and Its Discontents*. Harper & Row, New York.

Division of Elementary, Secondary, and Informal Education (1997). *Foundations: The Challenge and Promise of K-8 Science Education Reform*. National Science Foundation, Arlington, VA (NSF 97–76).

Division of Elementary, Secondary, and Informal Education (1999). *Foundations: Inquiry— Thoughts, Views, and Strategies for the K-5 Classroom*. National Science Foundation, Arlington, VA (NSF 99–148).

Donmoyer, R. (1998). Educational standards—moving beyond the rhetoric. *Educational Researcher*, 27(4):2, 10.

Doran, R., Chan, F., and Tamir, P. (1998). *Science Educator's Guide to Assessment*. National Science Teachers Association, Arlington, VA.

Education Week (1999). *Quality Counts '99: Rewarding Results, Punishing Failure*. Washington, DC.

Education Week (2000). *Quality Counts 2000: Who Should Teach?* Washington, DC.

Education Week (2001). *Quality Counts 2001. A Better Balance: Standards, Tests, and the Tools to Succeed*. Washington, DC.

Educational Testing Service (2000). *ETS Standards for Quality and Fairness*. Princeton, NJ.

Elman, B. A. (2000). *A Cultural History of Civil Examinations in Late Imperial China*. University of California Press, Berkeley, CA.

Elmore, R. F., and Rothman, R. (Eds.) (1999). *Testing, Teaching and Learning: A Guide for States and School Districts*. National Academy Press, Washington, DC.

Feuer, M. J., Holland, P. W., Green, B. F., Bertenthal, M. W., and Hemphill, F. C. (Eds.) (1999). *Uncommon Measures: Equivalence and Linkage Among Educational Tests*. National Academy Press, Washington, DC.

Fink, D. (2000). *Good Schools/Bad Schools: Why School Reform Doesn't Last*. Teachers College Press, New York.

Finn, C. E., Jr. (1991). *We Must Take Charge: Our Schools and Our Future*. Free Press, New York.

Firestone, W. A., Camilli, G., Yurecko, M., Monfils, L., and Mayrowetz, D. (July 26, 2000). State standards, socio-fiscal context and opportunity to learn in New Jersey. *Education Policy Analysis Archives*, 8(35), n.p. (*http://epaa.asu.edu/epaa/v8n35/*)

Fitzpatrick, A. R., and Yen, W. M. (1999). *Issues in Linking Scores on Alternative Assessments: Effects of Test Length and Sample Size on Test Reliability and Test Equating*. Council of Chief State School Officers, Washington, DC.

Forgione, P. (1998). Responses to frequently asked questions about 12th-grade TIMSS. *Phi Delta Kappan*, 79(10):769–772.

Fraknoi, A. (1999). Science education and the crisis of gullibility. In Chaisson, E. J. and Kim, T.-C. (Eds.), *The Thirteenth Labor: Improving Science Education*. Gordon and Breach Publishers, Newark, NJ, pp. 71–78.

Frechtling, J. (Ed.) (1993). *User-Friendly Handbook for Project Evaluation: Science, Mathematics, Engineering and Technology Education*. National Science Foundation, Arlington, VA (NSF 93–152).

Grant, C., and Graue, E. (1999). (Re)Viewing a review: A case history of the Review of Educational Research. *Review of Educational Research*, 69(4):384–396.

Horn, C., Ramos, M., Blumer, I., and Madaus, G. (2000). *Cut Scores: Results May Vary*. NBETP Monographs, Volume 1, Number 1. National Board on Educational Testing and Public Policy, Lynch School of Education, Boston College, Chestnut Hill, MA.

Hurd, P. D. (1997). *Inventing Science Education for the New Millennium*. Teachers College Press, New York.

International Technology Education Association (2000). *Standards for Technological Literacy: Content for the Study of Technology*. Reston, VA.

Jennings, J. F. (1998). *Why National Standards and Tests? Politics and the Quest for Better Schools*. Sage Publications, Thousand Oaks, CA.

Joint Committee on Standards for Educational Evaluation (1994). *The Program Evaluation Standards: How to Assess Evaluations of Educational Programs*. Sage Publications, Thousand Oaks, CA, 2nd ed.

Jones, L. V. (1996). A history of the National Assessment of Educational Progress and some questions about its future. *Educational Researcher*, 25(7):15–22.

Kline, S. J. (1985). What is technology? *Bulletin of Science, Technology and Society*, 5(3):215–218.

Knapp, M. S. (1997). Between systemic reforms and the mathematics and science classroom: The dynamics of innovation, implementation, and professional learning. *Review of Educational Research*, 67(2):227–266.

Krug, E. A. (1964). *The Shaping of the American High School. Volume 1*. University of Wisconsin Press, Madison, WI.

Krug, E. A. (1972). *The Shaping of the American High School. Volume 2*. University of Wisconsin Press, Madison, WI.

Kuhn, D. (1997). Constraints or guideposts? Developmental psychology and science education. *Review of Educational Research*, 67(1):141–150.

Kuhn, D. (1999). A developmental model of critical thinking. *Educational Researcher*, 28(2):16–25.

Lawrence, F., Huffman, D., and Welch, W. (2000). Policy considerations based on a cost analysis of alternative test formats in large-scale science assessments. *Journal of Research in Science Teaching*, 37(6):615–626.

Lee, O. (1999). Equity implications based on the conceptions of science achievement in major reform documents. *Review of Educational Research*, 69(1):83–115.

Leithwood, K., and Aitken, R. (1995). *Making Schools Smarter: A System for Monitoring School and District Progress*. Corwin Press, Thousand Oaks, CA.

Levesque, K., Bradby, D., Rossi, K., and Teitelbaum, P. (1998). *At Your Fingertips: Using Everyday Data to Improve Schools*. MPR Associates, Inc., Berkeley, CA.

Levin, H. M. (1998). Educational standards and the economy. *Educational Researcher*, 27(4):4–10.

Lindsay, G., and Desforges, M. (1998). *Baseline Assessment: Practice, Problems and Possibilities*. David Fulton Publishers, London.

Linn, M. C., and Muilenburg, L. (1996). Creating lifelong science learners: What models form a firm foundation? *Educational Researcher*, 25(5):18–24.

Loadman, W. E., and Thomas, A. M. (2000). Standardized test scores and alternative assessments: Different pieces of the same puzzle. *ENC Focus: Assessment that Informs Practice*, 7(2):18–20.

Madaus, G. (1990). *Testing as a Social Technology. The Inaugural Annual Boisi Lecture in Education and Public Policy*. Lynch School of Education, Boston College, Chestnut Hill, MA.

Martin, M. O., Mullis, I. V. S., Beaton, A. E., Gonzalez, E. J., Smith, T. A., and Kelly, D. A. (1997). *Science Achievement in the Primary School Years*. Lynch School of Education, Boston College, Chestnut Hill, MA.

Masters, G. N., and Keeves, J. P. (Eds.) (1999). *Advances in Measurement in Educational Research and Assessment*. Pergamon Press, New York.

Masters, G. N., Adams, R. J., and Wilson, M. (1999). Charting of student progress. In Masters, G. N. and Keeves, J. P. (Eds.), *Advances in Measurement in Educational Research and Assessment*. Pergamon Press, New York, pp. 254–267.

Mayer, D. P., Mullens, J. E., Moore, M. T., and Ralph, J. (2000). *Monitoring School Quality: An Indicators Report*. National Center for Education Statistics, U.S. Department of Education, Washington, DC.

McNeil, L. (2000). *Contradictions of Reform: Educational Costs of Standardized Testing*. Routledge, New York.

Metz, K. E. (1995). Reassessment of developmental constraints on children's science instruction. *Review of Educational Research*, 65(2):93–127.

Metz, K. E. (1997). On the complex relation between cognitive developmental research and children's science curricula. *Review of Educational Research*, 67(1):151–163.

Miller, D. W. (2001). Scholars say high-stakes tests deserve a failing grade. *The Chronicle of Higher Education*, March 2nd, pp. A14–A15.

Miller, M. D. (1999). *Teacher Uses and Perceptions of the Impact of Statewide Performance-Based Assessments*. Council of Chief State School Officers, Washington, DC.

Mintzes, J. J., Wandersee, J. H., and Novak, J. D. (Eds.) (1999). *Assessing Science Understanding: A Human Constructivist View*. Academic Press, New York.

Mohr, L. B. (1995). *Impact Analysis for Program Evaluation*. Sage Publications, Thousand Oaks, CA, 2nd ed.

Mote, F. W. (1999). *Imperial China, 900–1800*. Harvard University Press, Cambridge, MA.

Mullis, I. V. S., Martin, M. O., Beaton, A. E., Gonzalez, E. J., Kelly, D. L., and Smith, T. A. (1998). *Mathematics and Science Achievement in the Final Year of Secondary School*. Lynch School of Education, Boston College, Chestnut Hill, MA.

Mungazi, D. A. (1999). *The Evolution of Educational Theory in the United States*. Praeger Publishers, Westport, CT.

National Center for Education Statistics (1998). *Pursuing Excellence: A Study of Twelfth Grade Mathematics and Science Achievement in International Context*. U.S. Department of Education, Washington, DC.

National Commission on Excellence in Education (1983). *A Nation at Risk: The Imperative for Educational Reform*. U.S. Government Printing Office, Washington, DC.

National Council of Teachers of Mathematics (2000). *Principles and Standards for School Mathematics*. Reston, VA.

National Research Council (1996). *National Science Education Standards*. National Academy Press, Washington, DC.

National Research Council (1999). *High Stakes: Testing for Tracking, Promotion, and Graduation*. National Academy Press, Washington, DC.

North Central Regional Educational Laboratory (2000). *Blueprints: A Practical Toolkit for Designing and Facilitating Professional Development*. Eisenhower Regional Consortia for Mathematics and Science Education and Eisenhower National Clearinghouse, Oak Brook, IL.

Office of Educational Research and Improvement (1997). *Attaining Excellence. A TIMSS Resource Kit*. U.S. Department of Education, Washington, DC.

Olson, L. (2001). Usefulness of annual testing varies by state. *Education Week*, 20(23):1, 22, 23.

Organization for Economic Cooperation and Development (2000). *Education at a Glance*. Paris, France.

Osborne, J., and Pilot, A. (2000). Recent trends in the reform of science and technology curricula. Educational Innovation and Information, *International Bureau of Education*, UNESCO, June, Number 103, 2–7.

Parkes, J. (2000). The relationship between the reliability and cost of performance assessments. *Education Policy Analysis Archives*, 8(16):1–15. (online journal epaa.asu.edu/epaa/v8n16/)

Public Agenda (2001). Reality check 2001. *Education Week*, 20(23):S1–S8.

Research and Policy Committee (2001). *Measuring What Matters: Using Assessment and Accountability to Improve Student Learning*. Committee on Economic Development, New York.

Roach, J. (1971). *Public Examinations in England, 1850–1900*. Cambridge University Press, London.

Roberts, D. A. (1998). Analyzing school science courses: The concept of companion meaning. In Roberts, D. A. and Ostman, L. (Eds.), *Problems of Meaning in Science Curriculum*. Teachers College Press, New York, pp. 5–12.

Roseman, J. E. (2001). Personal communication with Project 2061, American Association for the Advancement of Science. Washington, DC.

Sanders, W. L., Saxton, A. M., and Horn, S. P. (1997). The Tennessee value-added assessment system; a quantitative, outcomes-based approach to educational measurement. In Millman, J. (Ed.), *Grading Teachers, Grading Schools. Is Student Achievement a Valid Evaluation Measure?* Corwin Press, Thousand Oaks, CA, pp. 137–162.

Sands, W. A., Waters, B. K., and McBride, J. R. (Eds.) (1997). *Computerized Adaptive Testing: From Inquiry to Operation*. American Psychological Association, Washington, DC.

Sarason, S. B. (1996). *Revisiting "The Culture of the School and the Problem of Change."* Teachers College Press, New York.

Schacter, J., Herl, H. E., Chung, G. K. W., Dennis, R. A., O'Neil, and H. F. Jr. (1999). Computer-based performance assessments: A solution to the narrow measurement and reporting of problem solving. *Computers in Human Behavior*, 15:403–418.

Schmidt, W. H., McKnight, C. C., and Raizen, S. A. (1997a). *A Splintered Vision: An Investigation of U.S. Science and Mathematics Education*. Kluwer Academic Publishers, Boston, MA.

Schmidt, W. H., Raizen, S. A., Britton, E. D., Bianchi, L. J., and Wolfe, R. G. (1997b). *Many Visions, Many Aims: A Cross-National Investigation of Curricular Intentions in School Science*. Kluwer Academic Publishers, Boston, MA.

Shadish, W. R., Jr., Cook, T. D., and Leviton, L. C. (1991). *Foundations of Program Evaluation: Theories of Practice*. Sage Publications, Thousand Oaks, CA.

Shepard, L. A., and Bliem, C. L. (1995). Parents' thinking about standardized tests and performance assessments. *Educational Researcher*, 24(8):25–32.

Skrla, L. (2001). Accountability, equity, and complexity. *Educational Researcher*, 30(4):15–21.

Spillane, J. P., and Callahan, K. A. (2000). Implementing state standards for science education: What district policy makers make of the hoopla. *Journal of Research in Science Teaching*, 37(5):401–425.

Stedman, L. C. (1997a). International achievement differences: An assessment of a new perspective. *Educational Researcher*, 26(3):4–15.

Stedman, L. C. (1997b). Response: Deep achievement problems: The case for reform still stands. *Educational Researcher*, 26(3):27–29.

Stine, M. A. (2000). State achievement tests can be a positive force in your classroom. *ENC Focus: Assessment that Informs Practice*, 7(2):45–46.

Tucker, M. S., and Codding, J. B. (1998). *Standards for Our Schools: How to Set Them, Measure Them, and Reach Them*. Jossey Bass, San Francisco.

Tyack, D., and Cuban, L. (1995). *Tinkering Toward Utopia: A Century of Public School Reform*. Harvard University Press, Cambridge, MA.

Tye, B. B. (2000). *Hard Truths: Uncovering the Deep Structure of Schooling*. Teachers College Press, New York.

Tymms, P. (1999). *Baseline Assessment and Monitoring in Primary Schools: Achievements, Attitudes, and Value-Added Indicators*. David Fulton Publishers, London.

Vinovskis, M. A. (1999). *History & Educational Policymaking*. Yale University Press, New Haven, CT.

Vinson, K. D., Gibson, R., and Ross, E. W. (2001). High-stakes testing and standardization: The threat to authenticity. *Progressive Perspectives*, 3(2):1–13.

Wahlstrom, D. (1999). *Using Data to Improve Student Achievement. A Handbook for Collecting, Organizing, Analyzing, and Using Data*. Successline, Inc., Virginia Beach, VA.

Webb, N. M., Schlackman, J., and Sugrue, B. (2000). *The Dependability and Interchangeability of Assessments Methods in Science*. CSE Technical Report 515. Center for the Study of Evaluation, UCLA, Los Angeles, CA.

Webb, N. L. (1999). *Alignment of Science and Mathematics Standards and Assessments in Four States*. National Institute for Science Education and Council of Chief State School Officers, Washington, DC.

Wilson, M., and Sloane, K. (2000). From principles to practice: An embedded assessment system. *Applied Measurement in Education*, 13(2):181–208.

Wolf, R. M. (1998). National standards: Do we need them? *Educational Researcher*, 27(4): 22–25.

Zeng, K. (1998). *Dragon Gate: Competitive Examinations and their Consequences*. Cassell Publishing, London.

CHAPTER 5

Assessment Reform
A Decade of Change

Wendy McColskey and Rita O'Sullivan

INTRODUCTION

The goal of this chapter is to provide a broad perspective on emerging issues
in student assessment for those engaged in implementing or evaluating
science curriculum reform efforts. Increasingly over the last half of the
1990s, state assessment programs have provided the measure of student
achievement that counts the most for many educators, policy-makers, and
program funders. Yet our experience in evaluating science curriculum
reform programs suggests to us that there are limitations in treating state
assessments as the most important measure of the quality of an instructional program. It is important that those in positions internal to school
systems with responsibilities for evaluating the success of instructional programs, and those who function as external evaluators evaluating program
impact, understand the issues in assessment reform. Such understanding
will lead to more informed choices about the design of student assessment

Wendy McColskey, Southeastern Regional Vision for Education, University of North Carolina
at Greensboro, P.O. Box 5367, Greensboro, NC 27435.   Rita O-Sullivan, School of Education, University of North Carolina, Chapel Hill, NC 27599.

Evaluation of Science and Technology Education at the Dawn of a New Millennium, edited by
James W. Altschuld and David D. Kumar, Kluwer Academic / Plenum Publishers, New York, 2002.

systems for program evaluation purposes and about the involvement of educators at different levels in the student assessment process.

A problem faced by educators and evaluators trying to use student assessment to improve programs and learning is that the extensive literature on assessment emanates from a variety of perspectives (e.g., psychometricians involved in test development; policy-makers trying to use large-scale assessment to drive improvements; educators who must respond to state accountability policies; researchers in measurement and psychology who are trying to understand the role assessment plays in learning; staff developers and consultants trying to help teachers implement better assessment practices in the classroom; curriculum developers trying to develop better assessments to fit curriculum goals; evaluators trying to design valid measures that will inform decisions about program improvement or funding, etc.). By way of an outline organized by levels of the educational system, we offer a perspective on a decade of lessons learned in the use of student assessment. We draw on our personal experiences in evaluating science curriculum reform programs and providing technical assistance, publications, and training to districts, schools, and teachers interested in improving the methods and practices of student assessment. We hope to leave the reader with a clearer picture of current and future assessment roles across the educational system (from state departments of education to teachers as evaluators of student learning in classrooms). Although science assessment is our reference point, the issues around how we assess student performance are not unique to science.

In the next section, we take a retrospective look at state assessment programs, as they remain a significant influence on the evaluation of instructional programs. Then, we summarize what we have learned about the roles of districts, schools, and teachers as regards their use of student assessment in the context of state and district standards. We conclude with implications for the future.

WHAT HAVE WE LEARNED ABOUT THE ROLE OF STATE ASSESSMENT PROGRAMS?

Statewide Assessment Programs, 1990–2000

Linn (2000) described five decades of influences related to large-scale testing of students. In the 1990s, state and federal assessment initiatives focused on: (a) the importance of aligning state tests with content and performance standards; (b) the inclusion of all students; and (c) finding ways to use test results to judge and label the effectiveness of schools (identify "failing" schools) and judge the proficiency of students (relative to promotion decisions). Linn concluded that significant problems surround using tests for school and student accountability purposes, but that policymakers are unlikely to back away from their attraction to accountability policies, which they believe will improve student outcomes. Curriculum reformers have been especially concerned about potential negative influences of high-stakes, large scale testing on the quality of instructional and assessment practices in classrooms.

> Results on standardized tests have been used as a yardstick to make judgments about what students know and can do. They also have been used to indicate how well or poorly a teacher, school, school system, or national education system is doing. In addition, they have functioned as gatekeepers, affecting school promotion, graduation, college entrance, acceptance into graduate and professional schools, and even employment opportunity. As test scores decline, the pressure to "improve test scores" causes elementary and secondary teachers to change the content of what they teach. They may spend significant instructional time in test preparation, omitting important content, because "it is not on the test." They will drill students to acquire factual knowledge, because "getting ready for the test" leaves them without time for in-depth exploration or the development of critical thinking by time-consuming means such as projects or hands-on activities. (Harmon, 1995, p. 32)

While still at issue in 2001, the concerns expressed in the quote above, about standardized testing and state assessment programs, reflect the anxiety of science curriculum reformers of the early 1990s who worried that testing practices of the time would not adequately capture the richness of the emerging constructivist approach to science education. These concerns led to a brief but intense effort mid-decade to develop and include alternative assessment procedures in state testing programs to create assessment opportunities for domains that included conceptual understanding, higher order thinking, and procedural knowledge. These efforts occurred as the quantity of standardized testing increased and states began implementing accountability policies searching for ways to enhance student achievement. By mid-decade, some states were in the process of developing new assessments to measure the outcomes of curriculum reform. Other states were busy revising assessment systems. A few states tried to engage school districts and teachers in site-based assessment reform.

Statewide Assessment Programs in the Early 1990s

According to "Surveying the Landscape of State Educational Assessment Programs" (Bond, Friedman, & Ploeg, 1993), the number of state assessment programs grew from 29 in 1980 to 46 by 1992. These assessment programs included nationally used, standardized achievement tests (e.g., California Achievement Tests, Iowa Tests of Basic Skills, etc.) as well as standardized tests developed to specifically reflect individual state curriculum guidelines. By 1993, Iowa, Nebraska, New Hampshire, and Wyoming were the only states without a statewide assessment program in any curriculum area. In 1992, of the 46 states with testing programs, 34 had science components. Only three of these states used some type of sampling strategy for their testing programs, as opposed to testing all students at a particular grade level or enrolled in a particular high school science course. Almost all of the states tested only three or four grade levels in science with North Carolina and Oklahoma noted exceptions, testing in more grades. Only 7 of the 34 states assessing science relied exclusively on norm-referenced standardized tests. The remaining 27 used criterion-referenced tests based on the state's curriculum; 4 of these 27 states used both norm-referenced and state curriculum-driven assessments. At the time, California, Kentucky, and New York used performance assessments as well as multiple-choice type standardized tests to assess their state's curriculum.

Statewide Assessment Systems Become Accountability Programs

By the year 2000, statewide assessment programs had evolved into standards-based accountability programs. As reported by Education Week (Orlofsky & Olson, 2001), all states now use testing programs to assess students' achievement. In addition, Orlofsky and Olson included, as an essential consideration, the extent to which states had clear and specific curriculum standards at elementary, middle, and high school, and the degree to which each state had accountability components in place. Assessment programs had gone beyond systematic collection of student achievement data to accountability programs that held schools responsible for student achievement. Accountability programs had provisions for school report cards, school ratings, rewards for good performance, assistance to schools in making improvements, and sanctions when schools consistently failed to meet standards.

By the end of the decade, the hopeful trend reported in 1993, that identified 17 states using performance assessments and 6 states using portfolios (Bond, Friedman, & Ploeg, 1993), succumbed to accountability, technical, professional development, and financial constraints. As of 2000, only seven states used extended response items in subjects other than English and only two states, Kentucky and Vermont, continued to use portfolios for statewide assessment purposes.

The Current Condition of Statewide Science Assessment

In comparison with 1993, a review of the 50 state assessment programs in science for the year 2000 (Education Week, 2001) revealed more testing based on state standards, but with a number of states reducing the number of grade levels where science was assessed or eliminating testing in science altogether. Nine states added assessments based on state standards, where none previously existed, and four added at least one level to the grades being assessed. Of the 34 states in 1993 that had some type of science assessment, 6 reduced one or more of the grade levels being tested, and 8 eliminated the science component of their assessment program entirely. Of the 21 states that currently do not assess science using state curriculum standards, 11 also did not test in science in 1993.

State Assessment Programs, Systemic Reform, and Curriculum Innovations

By the mid 1990s, science education in the United States was experiencing significant numbers of curriculum reform initiatives, and evaluators described a need for science educators to go beyond traditional multiple-choice type testing to assess these reform initiatives (e.g., Altschuld & Kumar, 1995; Harmon, 1995; Jeness & Barley, 1995; McColskey, Parke, Harmon, & Elliott, 1995; O'Sullivan, 1995; Veverka, 1995; Welch, 1995). At the same time, a number of studies were scrutinizing alternative assessments (e.g., Herman, Klein, Wakai, and Heath, 1995; Klein, McCaffrey, Stecher, Koretz, 1995; Linn, Burton, DeStefano, & Hanson, 1995; Pearlman, Tannenbaum, & Wesley, 1995) to determine their fairness and the utility of the format for large-scale testing purposes. Ultimately, the financial, systemic, and psychometric constraints surrounding alternative assessment in large scale testing precluded the growth optimistically predicted a few years earlier. The proliferation of high-stakes, state accountability programs, based primarily on multiple-choice type items, has resulted in concerns about inappropriate uses of test results and potential negative impacts on the quality of instruction (e.g., AERA, 2000; Kohn, 2001; Ohanian, 2001; Thompson, 2001). This debate and research on consequences of high-stakes testing programs will undoubtedly continue for the next few years as policymakers and researchers explore the effectiveness of accountability programs.

Trends in National Academic Progress in Mathematics and Science

A review of trends in academic progress from 1986 to 1999, throughout the United States, based on the National Assessment of Educational Progress (NAEP), revealed gains in mathematics and science achievement. These data are very useful as 1986 can be used as a benchmark year for the pre-curriculum reform efforts in science and mathematics. Curriculum reform efforts accelerated during the beginning of the 1990s and continued

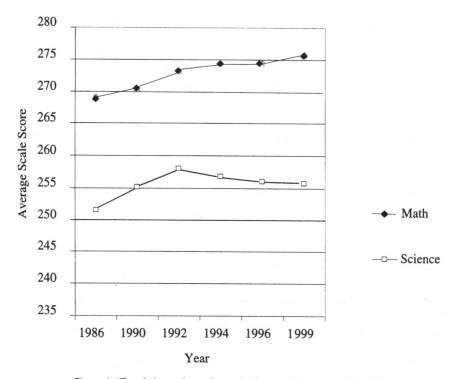

Figure 1. Trends in mathematics and science achievement, 1986–99.

at a slower pace on through the decade (as they competed with account-ability policies for teachers' attention). As can be seen from Figure 1, from 1986 to 1992, science and mathematics achievement both experienced upward trends, with the rate of science achievement outpacing that of math-ematics. We would hope that curriculum reform efforts in these two areas contributed in some way to these trends. But from 1992 to 1999, increases in science achievement held steady, while mathematics achievement con-tinued to rise. As previously noted, statewide assessment in science de-creased by 10 states from 1993 to 2000. Perhaps, some connection exists between the elimination of science testing in 20% of the states and the lack of progress on national science achievement assessments during this time.

Achievement in Science and Accountability Programs

Policymakers in the 1990s seem to have assumed that adoption of high stakes testing and accountability programs will result in improved achieve-ment. In a recent position statement concerning high-stakes testing in

PreK-12 education, the American Educational Research Association (2000) depicted the context of high stakes testing as "enacted by policy makers with the intention of improving education" (p. 1). The idea is that "setting high standards of achievement will inspire greater effort on the part of students, teachers, and educational administrators" (p. 2). Examining this widely accepted assumption is very important.

It is possible that even though achievement test scores may increase, the value of the increase in terms of true student progress or learning may be less than originally perceived. AERA (2000) warns:

> Policy makers and the public may be misled by spurious test score increases unrelated to any fundamental educational improvement; students may be placed at increased risk of educational failure and dropping out; teachers may be blamed or punished for inequitable resources over which they have no control, and curriculum and instruction may be severely distorted if high test scores per se, rather than learning become the overriding goal of classroom instruction. (p. 2)

Revisiting the National Assessment of Educational Progress (NAEP) science and mathematics scores presented earlier shows that the gains reported from 1986 to 1999 ranged from 251.4 to 255.8 for science and from 269 to 275.8 for mathematics. In the context of science education, the NAEP scale score level of 250 indicates that students can apply general scientific information. Roughly extrapolating the lines of progress into the future (on Figure 1), it appears that movement to an average score of 300, indicating that students are considered capable of analyzing scientific procedures and data, is unlikely without massive levels of support. That is, state testing programs, by themselves, seem unlikely to alter the current trajectory of these lines in a major way.

Another way to reflect on the contribution of state testing might be to look at achievement levels by state in science and compare the completeness of their accountability programs. Table 1 lists 19 of 40 states that participated in the 1996 NAEP, whose 8th grade students rated above the national average in science. The second column presents these states' 2000 accountability grades (out of 100) as assigned by *Education Week* (2001). These accountability grades were based on the use of state curriculum standards in core subject matter areas, types of assessments used in a state's assessment program, and number of accountability components that were operative. In looking for other indicators of science proficiency, the third column presents the available 1998 data on percentages of students enrolled in upper-level science courses and the fourth column reports whether that state had a standards-based science assessment program in place in 1992.

Although no systematic pattern is observable between science achievement and percentage of upper level science courses or presence of

Table 1. Student Achievement in Science, Accountability Grade (per Education Week 2001), Percentage of Students Enrolled in Upper Level Sciences Courses, and Presence of State Standards-Based Science Assessments in 1992

State	% in 1996 at or above proficient on NAEP science test	2000 Accountability Grades/100	% in 1996 in upper level science courses	1992 State Standards-Based Science— Assessment
Montana	41	40		YES
North Dakota	41	50	32	YES
Maine	41	76		YES
Wisconsin	39	65	37	YES
Minnesota	37	57	23	YES
Massachusetts	37	89	37	YES
Iowa	36	31	35	NO
Connecticut	36	81	33	YES
Nebraska	35	61	33	NO
Wyoming	34	60		NO
Vermont	34	65	29	NO
Colorado	32	86		YES
Utah	32	73	30	YES
Oregon	32	86	20	NO
Michigan	32	86	29	YES
Alaska	31	61		NO
Indiana	30	81	31	YES
Missouri	28	78	31	YES
Washington	27	67		YES

a statewide, standards-based assessment system in 1992, a relationship appears to exist between science proficiency level and accountability score. The relationship seems to suggest a possibility that states with more complete accountability programs tend to have lower science achievement as measured by NAEP. In fact, in analyzing the data for 40 states (for which data were available), science proficiency on NAEP correlated negatively with 2000 accountability grades ($r = -.37; p < .02$). Although some may interpret this to suggest that more complete accountability programs lead to lower achievement, it seems that states with lower achievement scores are more likely to institute accountability programs. And, of course, states with lower scores differ from higher performing states in a variety of other ways such as the incidence of poverty.

Do states most in need of improving achievement have complete statewide accountability programs? Table 2 reports the 11 states with the lowest ranks in terms of their 1996 NAEP science scores (range = 6%–16% at or above proficient). All have relatively high accountability grades, which

Table 2. Accountability Grades and NAEP Ranks for Eleven Lowest Ranked States

State	Rank in 1996 at or above science proficient on NAEP (40 states)	2000 Accountability Grades/100	Rank in 1996 at or above math proficient on NAEP (40 states)	Rank in 1992 at or above math proficient on NAEP (41 states)
West Virginia	30	69	34	37
Delaware	31.5	85	26	29
Florida	31.5	84	28	30
Georgia	33	78	31	34
California	34	85	29	27
New Mexico	35	91	35	36
Alabama	36	79	38	39
South Carolina	37	87	36	31
Hawaii	38	60	32	33
Louisiana	39	85	39	40
Mississippi	40	62	40	41

suggests that policy-makers in those states do have a high level of concern about student achievement. (Those 11 states with the lowest NAEP science achievement scores have an average accountability grade of C+ (78.6/100), while those ranked in the top 11 have a lower average accountability grade of D– (61.4/100).) In the last two columns, the ranks of the states in terms of mathematics achievement as reported by NAEP (science achievement data were not as readily accessible for this period) for 8th grade students in 1992 and 1996 show little change during the four-year period. In fact, 7 of the 10 lowest ranked states remain there across all three columns, which raises the question of whether a strong state accountability program, by itself, without a tightly coordinated and planned delivery of massive curricular reform, professional development, teacher development and recruitment, and other supporting policy initiatives aimed at all districts can be expected to increase student learning in a significant way.

WHAT HAVE WE LEARNED ABOUT ASSESSMENT REFORM AT THE DISTRICT, SCHOOL, AND CLASSROOM LEVEL?

In summarizing the last decade, states have implemented or revised content standards, testing programs, and accountability mechanisms as primary tools for raising student achievement. Some positive impacts of these state-level strategies have been recognized (e.g., increased attention and resources for low performing, high poverty schools). However, as mentioned above, we believe the ultimate success of this federal and state effort to improve educational outcomes will depend on much more than state accountability programs. It will depend on significant and coordinated efforts to build district, school, and teacher capacities to organize teaching and student assessment around a more challenging curriculum based on higher-order thinking skills and depth of content understanding (goals of many curriculum reforms).

Our personal experiences have convinced us that there is much work to do in getting student assessment right at these levels and evaluators can help educators develop a better understanding of the power of student assessment when used well. Treating the assessment of student performance as mostly a "state" issue is unlikely to ever get us toward the higher levels envisioned by the standards documents produced by the American Association for the Advancement of Science's Project 2061 or the National Research Council. We need clearer models of what standards-based assessment reform strategies look like at these levels to have a chance of convincing educators that they can stay focused on the curriculum revitalization envisioned by the standards documents and not resort to teaching to the objectives embodied in any particular state test. Below we summarize what we understand about standards-based assessment reform strategies at the district, school, and classroom.

The District Role

> As states construct their systems of standards-based reform, we must begin to examine how different policies, and combinations of policies, impact local policy and practice. More than twenty years of research on the implementation of state and federal education policies shows that change ultimately depends on the willingness and capacity of local communities, schools, and teachers to implement their policies. (Goertz, 2000, p. 66)

The district role in assessment reform is perhaps the least understood and researched of the three roles—district, school, and teacher. District leaders are gatekeepers and translators of state and federal policies and sense makers who try to communicate a sense of purpose to their schools based on what they are hearing and seeing from their various publics. The district role in standards and assessment is that of the first line of support to schools and should not mean just passing on state requirements to schools (e.g., "get your test scores up"). Taken together, a number of recent qualitative studies of the district role in reform provide some inkling of what the district role as the mediator or translator between state standards, assessment, and accountability policies and ultimately, teachers' classroom practices, might entail. These studies show district leaders taking an active role in:

1. Engaging their community in an understanding of state standards or district adaptations of state standards to meet local expectations, listening to the staff's and community's hopes and desires for students and using this information to set goals (Laboratory Network Program, 2000).

2. Articulating best practices expected of teachers (sometimes in the form of curriculum guides or model lessons or units, sometimes in the form of district policies, sometimes in the form of teacher evaluation systems).

3. Developing and implementing district-wide performance assessment systems that go beyond state tests and encourage teachers to teach higher-order thinking and develop deep content understanding (Hartenback *et al.*, 1997).

4. Thinking about graduation requirements that will prepare students for success after high school and implementing curriculum embedded assessments in high schools that allow students to demonstrate their ability to use key knowledge and skills meaningfully (graduation portfolios or exhibitions).

5. Providing expanded opportunities for professional growth, realizing that challenging goals are useless unless staff have the "know how" to better meet students' needs in the classroom (Laboratory Network Program, 2000); in some cases, providing "coaches" to help schools better understand the meaning of quality work for students.

6. Providing clear direction and feedback to principals about what is expected of them and in return, trying to decrease the distractions faced by schools in the form of mandates and requests that are extraneous to the goal of quality instruction in all classrooms.

7. Implementing district-wide mechanisms for coordination and ongoing continuous improvement across schools, such as K-12 teacher advisory groups in different discipline areas, summer meetings of school representatives to review progress in becoming standards-based, etc.

Although there is much being learned from "cutting edge" districts about how to translate standards, assessment, and accountability into a coherent set of strategies at the local level, one role that very few districts are taking on is monitoring the "conditions of instruction" (National

Research Council, 1999). That is, for curriculum reforms to be successful in the future, districts may need to get serious about monitoring not just test score results but the quality of what is going on in classrooms (and providing feedback and support to schools and teachers based on these data). A review from the National Research Council (1999) of the implementation status of standards, assessment, accountability, and professional development across states noted that because of a strong tendency for some schools to resort to short-term, quick fix responses to high stakes accountability policies (as opposed to long-term efforts to improve the quality of the instructional process), school districts must begin to monitor the quality of classroom practice across their schools and design interventions to build teacher capacity.

Corroborating this need for districts to monitor what goes on in classrooms, The National Center for Research on Evaluation, Standards, and Student Testing (CRESST) initiated a research project to design a tool to measure what is going on in classrooms. That is, there was a perceived need to go beyond test scores and collect data on not just what students have learned but the extent to which they have had meaningful opportunities to learn through the work assigned to them by teachers. Aschbacher (2000) reflected that "our experience in past studies at CRESST suggested that many teachers had difficulties maintaining high standards for student achievement and developing learning and assessment activities and grading criteria that are well aligned with student standards. Teachers' curriculum and instruction decisions tended to be driven by activities rather than by desired student outcomes, and the activities often emphasized participation rather than rigorous thinking or use of content knowledge" (p. 4).

CRESST is developing indicators of classroom practice so that progress in improving the quality of work provided to students can be tracked. This type of data, if used by districts and program evaluators, could be tremendously helpful in understanding the kinds of opportunities to learn students are receiving in a particular subject area. The CRESST Indicators are intended to

> Provide two valuable tools to help schools and districts enhance their capacity for improving education. The first is a set of indicators of classroom practice that provide an alternative to observation and teacher self-report. Such indicators could be used in research on teaching and learning, in large-scale evaluations, and in local self-evaluation efforts to monitor progress in instructional quality along with student performance over time. The second tool is a rubric with guidelines for describing the nature and quality of classroom assignments and linking them to student work. Just as looking at student work has become a popular and effective strategy for encouraging teachers to be more reflective about their classrooms, the rubric is meant to deepen and extend teachers' reflections on the quality of an assignment and its impact on the nature of student work. Such

a tool should be useful in both pre-service and in-service professional development. (p. 6)

Through sampling and rating teachers' assignments, data on the quality of classroom assignments are gathered. Data are reported on the cognitive challenge of the assignments, the clarity of goals for student learning articulated by the teachers, the clarity of grading criteria, the alignment of learning goals and task, the alignment of learning goals and grading criteria, and the overall quality of the assignment.

A second role that many districts have not had to play in the past is that of evaluating students' readiness to move on to the next level (to be promoted). At the same time, state policies that hold students individually accountable for achieving particular scores on state tests in order to be promoted are increasingly on the radar screen. That is, states are developing "student accountability" policies that imply that students must get passing scores on state tests to be promoted at certain grade levels. As a result, organizations like the American Educational Research Association and the American Psychological Association have issued statements warning against using a single state test result for the purpose of deciding if a student should be promoted. It appears that the burden of this warning that state tests should not be used as sole determinants of student promotion decisions will fall on districts. Yet developing systematic ways of evaluating students' readiness to move to the next level of schooling through the implementation of multiple measures of achievement (rather than a single state test score) is a very time-consuming process. This area may become a major challenge for districts in the immediate future.

In summary, our reflections on how "cutting edge" districts are translating the demand for higher standards into action and of new assessment roles emerging for districts demonstrate that assessment understanding is very important to a district's ability to make sense of a "high stakes" testing environment. In particular, it is important that assessment reform is seen as a much broader issue than just getting their schools to perform well on state tests. Some districts currently have very little capacity for implementing "assessment reform" as described. There is a clear need for states or other regional or national funding sources to help districts build their capacity for translating "high stakes" state test requirements into a meaningful kind of long-term focus on instructional improvement and evaluation of such improvement at the classroom level.

The School Role

School leaders are next in line to react to states standards, assessment, and accountability policy. A recent exploratory study of nine schools in

Michigan (NCREL, 2000) identified several strategies that perhaps begin to define what "assessment reform" looks like at the school level:

1. Schools are strategic in their choices and uses of formal assessments and have a deliberately organized set of assessment tools used for clearly articulated purposes. That is, they look at the assessments used in the school to inform them of students' progress and consider how they fit together and lead toward a common vision.

2. Schools work to align standards with curriculum, instruction, and assessment. In other words, schools have tools such as "curriculum mapping" (comparison of enacted instructional goals—what is being taught—to state or district standards adopted by the school) to identify gaps and repetitions and to make the instructional program more focused and coherent across teachers in the school.

3. Schools provide support to teachers and staff in developing their understanding and skills in student assessment realizing that this is an area in which many teachers feel unprepared and that the quality of work assigned to students has implications for how well they learn.

4. Schools find time to involve teachers in assessment related activities so that a "common language" develops. The activities might include developing, piloting, and revising performance-based assessments; looking at test data from various sources; and looking at student work samples and asking questions of each other about it.

5. Schools realize that there are different purposes for assessment data; some data are useful for monitoring student progress (e.g., portfolios), some are useful for setting standards (e.g., graduation exhibitions), some are useful for program evaluations (performance tests reflecting the goals of a new science program), and some are useful for accountability (e.g., the percent of students below grade level on state tests). Schools must evaluate their mix of assessment data to see if it fits their needs.

The NCREL report notes that most educators have received little formal training in student assessment and the use of data to improve teaching and learning. At the same time, the task of making sense of state standards, assessment, and accountability policies in a way that supports a coherent rather than fragmented or extremely narrowed school vision of student learning and achievement will take much time, attention, and energy. The implication is that "assessment reform" at the school level is unlikely to bubble up as a school-based initiative. States and districts will need to create the support infrastructure (incentives and guidance) for schools to understand the role of assessment in improving learning and take on the hard work of curriculum mapping, analyzing assessment uses, developing performance-based assessments when needed, monitoring student progress, looking at student work samples, and making sense of accountability demands.

The logic behind standards-based reform (including attention to alternative assessment) is that schools have tended to overemphasize what is easiest to assess (factual knowledge that can be assessed by objective tests) and have underemphasized the critical thinking and reasoning skills that are harder to assess but which are so important for functioning in a complex society. That is, the goal of standards-based reform is in part to change the nature of what goes on in schools such that learning involves more than just rote kinds of memorization activities.

> That the common curriculum should address challenging standards aimed at higher-order thinking and problem solving is likewise a rejection of past practices and theory. The transmission model of learning based on rote memorization of isolated facts removed learning from contexts that could provide both meaning and application. By watering down curricula and emphasizing minimum competencies, schools have lowered expectations and limited opportunities to learn. By contrast, if children are presented with more challenging and complex problems and given the support to solve them, they will develop deeper understandings and at the same time become more adept at the modes of inquiry and ways of reasoning that will help them solve new problems in the future. (Shepard, 2000, p. 14)

There is evidence that schools can be successful in "assessment reform", particularly in implementing alternative approaches to assessment. *In Authentic Assessment in Action*, Darling-Hammond, Ancess, and Falk (1995) documented the work of three high schools and two elementary schools engaged in trying to move beyond traditional grades and standardized tests to more "authentic" looks at what students could do. These schools developed ways to focus students on challenging, performance-based tasks that required analysis, integration of knowledge, and invention, as well as written and oral expression of ideas and concepts.

The term "authentic assessment" was introduced in the early 1990s as a way of suggesting that the kind of work students do in schools should include some tasks that have characteristics in common with the kind of work done in that discipline in the "real-world." The five schools described in the Darling-Hammond *et al.* (1995) book were committed to assessing student performance through exhibitions of real performances rather than through traditional paper and pencil testing methods. The authors concluded that the school case studies "illustrate how powerfully teachers and principals can use assessment practices as a means to help themselves become more successful with students and more accountable for learning— both their ongoing learning and the learning of their students" (p. 252).

Effective school-wide use of performance-based or alternative assessment depended on several factors, according to Darling-Hammond *et al.*: (a) Schools must function as communities and "deprivatize" teaching such that expectations for students are clear and common instructional goals inform teachers work, (b) Schools must reorganize to enable teachers to work together to develop, implement, and assess critical student performances, and (c) Faculty must be deeply involved in the development, implementation, and scoring of critical student performances. "Assessments that are externally developed and scored cannot transform the knowledge and understandings of teachers—and of school organizations—even if they are more performance-based than are current tests" (p. 253).

As a result of developing and using such assessments, teachers in these schools became more reflective about goals, their teaching strategies and learning environments. They also were better able to understand each student's strengths, weaknesses, learning styles, interests, etc. and thus, better respond to them individually as learners. It is clear that schools can initiate assessment reform under the right kind of leadership and with the right kinds of support. (These schools were all members of the Coalition of Essential Schools network, which provided some of the "reform" ideas.)

In summary, we suggest that some exceptional schools have understood the power of assessment in framing for students a clearer view of their learning targets and have designed performance assessments that ask students to take on a significant learning experience and share the results of their learning experience publicly in forums arranged by the school. Developing and implementing school-wide performance assessments (whether it is in the form of performances or exhibitions, portfolios, or student-led conferencing) take a great deal of faculty time and energy. It is not likely to become common practice unless there are concentrated efforts by states and districts to provide the resources and expertise needed.

The Teacher Role

Ultimately, our success in achieving higher standards comes down to what teachers can do differently in their classrooms to help more students become confident and successful learners. Black and Wiliam (1998) called this the "black box" of the classroom in that policy makers deal with inputs (standards, high-stakes tests, funding, rules, etc.) and demand certain outputs (test results) but have focused relatively little effort on direct help and support to teachers to rethink how their classroom practices might need to change in light of the demand for improved student performance. Black and Wiliam also suggest that better assessment in the classroom can help raise standards. Similarly, Stiggins (1999) has argued that even though large-scale student assessment is where states put significant dollars, it is the teacher's use of assessment in the classroom that truly determines the depth of knowledge and skills that students develop over time. For example, if all students see from their science teacher is multiple-choice and short-answer factual knowledge questions, assuming that instruction is also geared towards factual knowledge, they are likely to be less able to explain important concepts, develop ways to test out hypotheses, and apply knowledge than students in classrooms with teachers who consistently challenge them to perform in these ways.

How might the "black box" of the classroom need to change if higher standards are to become a reality? Those who have worked closely with teachers in thinking about "good" classroom assessment practices argue that, in general, assessment geared toward producing high levels of learning across all levels of ability is not happening in many classrooms. Black and Wiliam (1998) summarize some of the problems in classroom assessment as:

1. Classroom assessment practices generally focus more on superficial or rote learning, concentrating on recall of isolated details, which students soon forget.
2. Teachers do not review their assessments/assignments or get peers to review them so there is little critical reflection on what is being assessed and why.
3. The grading aspect of assessment is overemphasized and the learning or improvement purpose of assessment is underemphasized.

Shepard (2000) has argued that changing classroom assessment practices involves:

1. Changing the nature of the assessment conversations teachers have with students such that students develop greater knowledge of and responsibility for learning goals.
2. Assessing students' prior knowledge and using that information in planning better instruction to meet their needs and match their interests.

3. Giving students feedback in ways that go beyond grades such that they are helped to understand what quality work or thinking looks like.
4. Getting clearer about the explicit criteria for open-ended/ performance tasks and involving students in self-assessing.
5. Using information from students to evaluate and improve their teaching strategies.

Actualizing this kind of assessment reform at the classroom and school level is a long-term endeavor. To be successful in bringing along more students to higher levels of understanding, there will need to be deep changes in teachers' choices of assessments (kinds of work students are asked to do). That is, too many students, and perhaps more often students in economically disadvantaged communities, receive low-level, less than challenging work assignments (Puma *et al.*, 1997). Black and Wiliam concluded that "both in questioning and in written work, teachers' assessment focuses on low-level aims, mainly recall. There is little focus on such outcomes as speculation and critical reflection" (p. 9). Similarly, a study by Aschbacher (2000) found that the majority of classroom assignments sampled (in reading, writing, and project assignments) in third and seventh grade urban school classrooms were not "intellectually challenging" (tasks did not ask students to use higher order thinking).

A major factor contributing to the problems described above is teacher isolation. Teachers traditionally have operated independently such that many do not have opportunities to discuss instructional goals and the quality of their assignments with peers in any regular, structured kind of way. Thus, teachers may think they are teaching to and assessing high levels of student understanding when, in reality, they are not. One study of biology teachers' assessment practices showed that although teachers reported broad thinking and problem-solving goals for students, their assessment practices represented recognition and recall kinds of skills (Bol & Strage, 1996). One suggestion made by the authors was that teachers be encouraged to reflect on the numbers of higher-level, open-ended, or application types of items on their tests.

> One way to meet the demands for curricular reform in science education is for teachers to evaluate the correspondence between their instructional goals and assessment outcomes. ... Teachers also need to recognize that students are likely to actively resist more cognitively complex tasks for which they are held accountable, and this resistance may lead the teachers to stray from their initial instructional goals. A strategy for meeting the challenge to reform science classrooms is to enhance teachers' understanding of the importance of aligning their instructional goals with their tests and with the practice opportunities that support their assessment practices. (Bol & Strage, 1996, p. 161)

Even in college classrooms, those who have studied assessment practices have found that teachers tend to think they are teaching to higher-order thinking goals, that when student assignments are examined, tend to be weakly represented (Angelo & Cross, 1993).

In light of the classroom assessment problems identified above, what do we know about how to support reform at this level? Below we offer, not a comprehensive review of research and practice in the area of teachers' use of assessment in the service of curriculum reform, but a sample of important studies that contribute to our understanding of where we need to concentrate future assessment efforts in science education.

Embedding comprehensive assessment systems into curriculum materials and providing teachers support. Wilson and Sloane (2000) described an assessment system they developed to help teachers assess, interpret, and monitor student performance for a course within a middle school science curriculum (The Science Education for Public Understanding Project—SEPUP). Three primary teaching goals or student progress variables were identified as organizers for the assessment system (Understanding Concepts, Designing and Conducting Investigations, and Evidence and Trade-offs). Each progress variable has sub-objectives. Consistent with the 1990s standards movement argument for greater use of alternative or performance assessment as a way to improve students' problem-solving and higher-order thinking skills, assessment tasks were all open-ended and required student explanations. Recognizing that open-ended responses require substantial scoring time, teachers were provided scoring rubrics based on the progress variables. They also received "exemplars" that provided student work examples at the various levels of the scoring rubric. Additionally, they received a blueprint showing where assessment tasks should be given during the year.

An evaluation of the implementation of the assessment system was conducted using 63 teachers. Of these, 26 teachers had professional development in the use of the assessment system which included participation in training sessions and more importantly, participation in "moderation" sessions in which teachers discussed their students' scores on assessment tasks, making sure that scores were assigned consistently across teachers. These moderation sessions also allowed opportunities for teachers to discuss the implications of scores for future instruction (e.g., how to explain difficult concepts or address common student mistakes). Another 25 teachers used the assessment system but without direct training or participation with other teachers in looking at student work. A group of 12 teachers who taught the regular middle school science curriculum served as a comparison group.

Wilson and Sloane found that students with teachers who had training in the assessment system *and who were involved in ongoing collaborative work sessions to score and examine student work* gained significantly more from pretest to posttest than did the students with teachers who had the same assessment materials (assessment tasks and progress rubrics) but did not participate in the teacher discussion groups. The students with teachers in the comparison group (who taught the regular middle school science curriculum) had pre to post test gains similar to those with teachers who had the assessment materials but who didn't have the additional assessment training and group discussion opportunities.

Wilson and Sloane concluded that just providing the SEPUP curriculum and assessment materials to teachers was not enough to get educationally significant learning results (when compared to a control group). However, when teachers using the SEPUP program were provided with professional development opportunities around the use of the assessment system (teachers met regularly to examine the student work produced on the common assessments), the evaluation showed educationally significant student growth (compared to the non-assessment trained and comparison teachers).

This study has significant implications for the kinds of support teachers need. It is not just the assessment tasks they need, but the support and ongoing structured conversations in what their students' work on the assessment tasks tells them about how to change, target, or improve their subsequent instruction. Textbooks and "innovative" curriculum programs have routinely provided tests and sample assignments as part of the materials provided to teachers. This evaluation suggested that perhaps the most critical piece of support is opportunities for teachers to reflect on what student results mean. The insights gained from such group discussions lead to more focused and effective learning opportunities for students.

Making learning expectations explicit via rubrics and exemplars so that all students, and particularly low-achieving students, learn how to produce quality responses to assignments. White and Frederiksen (1998) started from the premise that high-level science learning opportunities are not made accessible to students of differing abilities. They set out to develop a classroom intervention that would make science literacy skills more accessible to traditionally low-achieving students. They designed an intervention to use in 12 urban science classes. The part of this study relevant to assessment had to do with a "Reflective-Assessment Process" that was introduced in some participating teachers' classes but not others. The authors developed a generic rubric for "Judging Research" that all 12 teachers used in judging student work on an assigned Research Project.

At the beginning of the unit, the "Judging Research" rubric was introduced to students who then used these criteria in completing their research projects. When students presented their research projects to the class, they also used this rubric to give each other feedback, both written and verbal. Each participating teacher used the Reflective-Assessment (RA) process in one half of his or her classes. These RA classes had in-class discussions about the quality of their research process and products. The control classes were introduced to the "Judging Research" rubric but did not have the self and peer assessment discussions around their project work.

The study found that on the Research Project assignment, the students receiving the Reflective-Assessment discussions did significantly better than the control students and the effect was greatest for low-achieving students. That is, the mean difference in the project scores between high and low achieving students (as determined by previous standardized tests) was significant in control classes but non-existent in the Reflective-Assessment classes. In this study, it seemed it was not a rubric by itself that made the difference (as teachers used the rubric in all the classes) but the rubric in the context of teacher-led discussions that involved student self and peer assessment. White and Frederiksen offered the following observations:

1. Learning to use assessment criteria matters (students who learned to use the scoring rubric well as a self-assessment tool had higher scores on projects than those who didn't learn to use the rubric well).
2. Reflective assessment increases student motivation (students in RA classes turned in projects at a higher rate and low-achieving students in RA classes more frequently chose difficult topics to research).

This research study extends the findings of the Wilson and Sloane study cited previously, where we learned that providing teachers with an assessment system to use in a course they teach is a particularly powerful intervention when teachers are given collaborative and structured opportunities to discuss the work students produce. Similarly, when teachers involved students in discussions about their work on a project (guided by clear assessment criteria) and how their work compared to the work produced by others, students seemed to develop insights about how their work could be improved and a better understanding of what "good" work was in ways that went beyond what they could understand from access to the rubric alone.

Providing teachers with in-depth professional development opportunities to become assessment literate. Stiggins, Director of the Assessment Training Institute, who has worked extensively in the area of professional development for educators in assessment literacy, has most recently focused his efforts on statewide initiatives to support teachers in developing their assessment knowledge and skills. Those involved over the last decade with classroom assessment realize that becoming "assessment literate" is not likely to happen as a result of a one-shot in-service workshop. It involves experimentation with different kinds of assessments, rubrics, and forms of feedback, sharing results from experimentation with someone who can challenge thinking a little further, then more reading, more experimentation, etc. It is a long-term, iterative process of reviewing your assessment system and teaching goals and tightening up the alignment. Learning about assessment and improving assessment practices are difficult to do in isolation. Stiggins, in collaboration with several states, is offering opportunities for schools to create teacher-learning groups as a vehicle for teachers to make the necessary long-term commitment to becoming more assessment literate. He suggests that professional development must provide for

1. An infusion of new ideas regarding effective assessment practices.
2. Opportunities to practice applying the principles of sound assessment.
3. Assessments that deliver concrete benefits almost immediately.
4. Ways for educators to take responsibility for their own training.
5. Flexibility (of schedules and resources).

He suggested that teachers' learning about assessment should use three strategies (10% workshops; 25% in learning team meetings; and 65% in individual study, classroom experimentation, and reflection on what works in the classroom). Districts and schools need to be prepared to pay for the costs of learning team meetings (substitutes, early release time, extended contract time) if they are serious about building teacher capacity in this area. The Assessment Institute offers publications and products to support teachers involved in learning more about assessment in collaborative group settings (Arter & Busick, 2001; Stiggins, 2001).

While not an exhaustive list of strategies, this section identified three strategies to build teachers' assessment skills emerging from research and practice. In summary, those working directly with teachers might consider:

1. Providing teachers who teach like-courses (e.g., 7th grade science) with facilitated or structured opportunities to discuss the quality of work their students are producing on key assessment tasks related to critical course objectives. Teachers should be involved in looking at student work relative to key course outcomes (students' progress in designing and conducting

      investigations, communicating scientific information, under-
      standing of concepts, etc.).

2.   Involving teachers in efforts to make their expectations or cri-
      teria for good work on open-ended or alternative assessment
      tasks crystal clear to all students, especially traditionally low-
      achieving students, and to involve their students in "reflective
      assessment" (self and peer assessment of work) around those
      criteria.

3.   Providing incentives and support to teachers interested in par-
      ticipating in teacher learning teams so that they can work with
      peers to implement and learn about "assessment reform" at the
      classroom level and get feedback and advice about the changes
      they try to make.

WHAT IS THE FUTURE OF ASSESSMENT REFORM?

Many of the early nineties conversations about the higher levels of thinking and problem-solving that students would need to be successful in a
highly technological and changing world of work were captured in national
and state standards documents. But then as state testing programs came on
board and increasingly were used for accountability, some educators began
to focus on the operationalized vision as found in the test items themselves
rather than the larger vision of what students should know and be able
to do as expressed in standards documents. Standards documents have
articulated the kinds of thinking that students need to develop but more
recently, state tests have seemed to become the guiding vision for some educators rather than the more comprehensive vision found in the standards.

    Even state departments are concerned about schools teaching to the
short-term outcomes (the state tests) to the exclusion of outcomes considered important but not tested by state tests, as evidenced by the following
excerpt:

> It is easy to understand why the high stakes consequences of our ABCs Account-
> ability Model lead many school staff to believe that it is essential to focus on
> assessments that resemble state assessments. The thinking goes, 'If we do not
> teach and test in the classroom the way the state tests, we will not make expected
> growth on ABCs'. However, this thinking limits the opportunities for students,
> the likelihood of mastering the North Carolina Standard Course of Study goals
> and objectives, and the rich possibilities for using assessment as a teaching tool.
> While schools may well achieve 'expected' or 'exemplary' growth by narrowing
> the teaching and testing focus to resemble state assessments, this approach is
> shortsighted. In addition, there are many schools making 'exemplary' growth
> that are not limiting teaching and learning in this way. The North Carolina

Department of Public Instruction supports the continued development of
quality teaching and classroom assessment as vehicles to prepare students to
master rigorous content and performance standards as well as do well on
accountability measures. We believe that the strategies most likely to result in
long-term growth and learning of high quality will result from effective use of
quality classroom assessments as an integral part of instruction. Additionally,
strong classroom assessment engages students in self-assessment and greater
ownership of their own learning. (North Carolina Department of Public Instruction, 1999, p. x)

In the last decade, we have seen somewhat competing goals traded off
in our state policies (the goal embodied in the standards documents versus
the goal embodied in accountability policies that identify the lowest achieving schools). The implicit goal of standards documents (particularly those
at the national level) is curriculum reform. In science, reform documents
such as Benchmarks for Science Literacy by the American Association
for the Advancement of Science's Project 2061 and the National Science
Education Standards developed by the National Research Council have
described a vision in which students have "the opportunity to learn fundamental concepts in depth, to develop subject matter knowledge in the
context of inquiry, and to become adept at using scientific knowledge to
address societal issues and make personal decisions" (p. 20, Shepard, 2000).
It is easy to see how the skills outlined in such documents (being able to
formulate questions, to design and carry out scientific investigations, to use
data collection tools, to develop and then defend a scientific argument, to
evaluate the evidence for a particular explanation, and to communicate the
results of research) are difficult to assess efficiently with large-scale tests.

In our opinion, a second goal has become primary in the late 1990s as
state accountability programs were implemented. The goal is not explicitly
curriculum reform, but the desire to hold all schools, especially those at the
bottom in terms of standardized test scores, accountable for moving more
and more students toward "grade level" functioning on basic math and
reading/writing skills. The introduction of this goal, although important, has
unfortunately seemed to supplant rather than supplement the goal of curriculum reform at the district, school, and in particular, at the classroom
level. The labeling or grading of schools based primarily on test scores has
tended to drive out conversations about the quality of the curriculum and
the kinds of abilities students need to be scientifically literate (or prepared
for the "real world"). Conversations in schools in 2001 are likely to revolve
around what is needed to help students do well on the "state test".

How might we begin to reestablish a balance of the two goals of
curriculum reform and accountability, in particular as they impact student-
learning opportunities in classrooms? Gipps (1999) argued that regardless
of the future of external, state-driven, "high-stakes" testing, it seemed clear

that the key for the future lies in the development of better assessment at the classroom and school level, where the learning is happening. To become self-initiating learners to the extent now needed in a rapidly changing world, students can't just be passive recipients of grades and test scores handed down to them. They have to understand the desired instructional goals and standards of work (criteria for what quality work means) so that they can self-assess and become progressively independent in improving the quality of the schoolwork they produce. In using the term "assessment reform" in this chapter, we are agreeing that for students to begin to perform at higher levels, especially those not traditionally successful in school, the power relationship between teacher and student will need to change such that students are brought into the assessment process, particularly around what it means to do "quality" work and why it is important.

For those states and funding organizations truly interested in the curriculum reform described in many national and state standards documents, perhaps the best point of entry or leverage for the future is local districts. For curriculum reform, which includes assessment reform (clearer and more focused instructional goals, better assessment tasks, better rubrics, better self and peer assessment, better formative feedback, etc.) to be taken seriously at the classroom level, the agents of change that must be engaged are local district leaders. It is very difficult to work with teachers and/or schools individually if the intent is to impact practice in a sustained and systemic way. Districts need guidance and encouragement to begin to develop strategies for adult learning that will result in curriculum and assessment reform on a broad scale. They need help in beginning to evaluate classroom quality.

Just as research and experience is demonstrating the power of providing structure and opportunities for teachers to work together in groups to critically look at the evidence that their students are achieving the desired instructional goals, perhaps it is time for states and others to provide structure and opportunities for districts to be systematically engaged in critically examining the evidence that their instructional programs in particular areas (e.g., science) are achieving the desired instructional goals and if not, why not.

REFERENCES

Altschuld, J. W., and Kumar, D. D. (1995). Program evaluation in science education: The model perspective. In O'Sullivan, R. G. (Ed.), Emerging roles of evaluation in reforming science education [Special issue]. *New Directions for Program Evaluation* 65: pp. 5–18.

American Association for the Advancement of Science. (1993). *Benchmarks for Science Literacy*. Oxford University Press, New York.

American Educational Research Association. (2000). *AERA Position Statement Concerning High-Stakes Testing in PreK-12 Education*. Author, Washington, DC.

Angelo, T., and Cross, K. P. (1993). *Classroom Assessment Techniques: A Handbook for College Teachers*. Jossey-Bass, San Francisco, CA.

Arter, J. A., and Busick, K. U. (2001). *Practice With Student-involved Classroom Assessment*. Assessment Training Institute, Portland, OR.

Aschbacher, P. R. (2000). Developing indicators of classroom practice to monitor and support school reform. *The CRESST Line* Winter:6–8.

Black, P., and Wiliam, D. (1998). Assessment and classroom learning. *Assessment in Education: Principles, Policy, and Practice* 5(1):7–74.

Black, P., and Wiliam, D. (1998). Inside the black box: Raising standards through classroom assessment. *Phi Delta Kappan* October:139–148.

Bol, L., and Strage, A. (1996). The contradiction between teachers' instructional goals and their assessment practices in high school biology courses. *Science Education* 80(2):145–163.

Bond, L., Friedman, L., and Ploeg, A. (Eds.). (1993). *Surveying the Landscape of State Educational Assessment Programs*. The Council for Educational Development and Research, Washington, DC.

Darling-Hammond, L., Ancess, J., and Falk, B. (1995). *Authentic Assessment in Action: Studies of Schools and Students at Work*. Teachers College Press, New York.

Gipps, C. V. (1999). Socio-cultural aspects of assessment. In Iran-Nejad, A. and Pearson, P. D. (Eds.), *Review of Research in Education* (Vol. 24). American Educational Research Association, Washington, DC, pp. 355–392.

Goertz, M. E. (2000). Implementing standards-based reform: Challenges for state policy. In Duggan, T. and Holmes, M. (Eds.), *Closing The Gap: A Report on The Wingspread Conference, Beyond the Standards Horse Race: Implementation, Assessment, and Accountability—The Keys to Improving Student Achievement*. Council for Basic Education, Washington, DC, pp. 65–83.

Harmon, M. (1995). The changing role of assessment in evaluating science education reform. In O'Sullivan, R. G. (Ed.), Emerging roles of evaluation in reforming science education [Special issue]. *New Directions for Program Evaluation* 65:31–52.

Hartenback, D. L., Ott, J., and Clark, S. (1997). Performance-based education in Aurora. *Educational Leadership* 51–55.

Herman, J. L., Klein, D. C. D., Wakai, S., and Heath, T. (1995, April). Assessing equity in alternative assessment: An illustration of opportunity-to-learn issues. In *Equity & Fairness in Large-scale Performance Assessment: What do the data show?* Symposium conducted at the annual meeting of the American Educational Research Association, University of California, Los Angeles, Los Angeles, CA.

Jenness, M. J., and Barley, Z. A. (1995). Using cluster evaluation in the context of science education reform. In O'Sullivan, R. G. (Ed.), Emerging roles of evaluation in reforming science education [Special issue]. *New Directions for Program Evaluation* 65:53–70.

Klein, S. D., McCaffrey, B., Stecher, and Koretz, D. (Unpublished manuscript). *The Reliability of Mathematics Portfolio Scores: Lessons from the Vermont Experience*.

Kohn, Ohanian, and Thompson. (2000). *District Approaches to Education Reform*. Mid-continent Research for Education and Learning, Aurora, CO.

Linn, R. L., Burton, E., DeStefano, L., and Hanson, M. (Unpublished manuscript). *Generalizability of New Standards Project 1993 Pilot Study Tasks in Mathematics*.

McColskey, W., Parke, H. M., Harman, P., and Elliott, R. M. (1995). Evaluations as collaborators in science education reform. In O'Sullivan, R. G. (Ed.), Emerging roles of evaluation in reforming science education [Special issue]. *New Directions for Program Evaluation* 65:71–89.

National Assessment of Educational Progress. (1999). *Trends in Academic Progress.* Author, Washington, DC.

National Center for Research on Evaluation, Standards, and Student Testing. (1999). *Developing Indicators of Classroom Practice to Monitor and Support School Reform* (Tech. Rep. No. 513). University of California, Center for the Study of Evaluation, Graduate School of Education and Information Studies, Los Angeles, CA.

National Research Council. (1996). *National Science Education Standards.* National Academy of Sciences, Washington, DC.

National Research Council. (1999). *Teaching, Testing, and Learning: A Guide for States and School Districts.* National Academy Press Board on Testing and Assessment, Commission on Behavioral and Social Science, and Education, Committee on Title I Testing and Assessment, Washington, DC.

North Carolina Department of Public Instruction. (1999). *Classroom Assessment: Linking Instruction and Assessment.* Author, Raleigh, NC.

North Central Regional Educational Laboratory. (2000*). Using Student Assessment Data: What Can We Learn from Schools?* (Policy Issue No. 6). Author, Chicago, IL.

O'Sullivan, R. G. (1995). From judges to collaborators: Evaluators' roles in science curriculum reform. In O'Sullivan, R. G. (Ed.), Emerging roles of evaluation in reforming science education [Special issue]. *New Directions for Program Evaluation* 65:10–30.

Orlofsky, G. F., and Olson, L. (2001, January 11). The state of the states. *Quality counts 2000* (Vol. 20, No. 17, pp. 86–88). Education Week: Washington, DC.

Pearlman, M. R., Tannenbaum, and Wesley, S. (1995, April). *Operational Implications of Development: The Paradox of Performance Assessment.* Paper presented at the annual meeting of the American Educational Research Association, San Francisco, CA.

Puma, M. J., Karweit, N., Price, C., Ricciati, A., Thompson, W., and Vaden-Kiernan, M. (1997). *Prospects: Final Report on Student Outcomes.* ABT Associates, Cambridge, MA.

Shepard, L. A. (2000). *The role of classroom assessment in teaching and learning.* Center for Research on Education, Diversity, and Excellence, Washington, DC.

Stiggins, R. (1999). Assessment, student confidence, and school success. *Phi Delta Kappan* November:191–198.

Stiggins, R. (2001). *Student-Involved Classroom Assessment* (3rd ed.). Merrill/Prentice Hall, Columbus, OH.

Veverka, K. D. (1995). *Fifth Grade Students' Perceptions about Group Science Assessment: An Exploratory Study.* Paper presented at annual meeting of the America Educational Research Association. University of Kansas.

Welch, W. N. (1995). Student assessment and curriculum evaluation. In Fraser, J. B. and Walberg, H. J. (Eds.), *Improving Science Education.* The National Society for the Study of Education, Chicago, IL, pp. 90–116.

White, B. Y., and Frederiksen, J. R. (1998). Inquiry, modeling, and metacognition: Making science accessible to all students. *Cognition and Instruction*, 16(1):3–118.

Wilson, M., and Sloane, K. (2000). From principles to practice: An embedded assessment system. *Applied Measurement in Education* 13(2):181–208.

CHAPTER 6

Evaluation of Information Technology

John M. Owen, Gerard T. Calnin, and Faye C. Lambert

INTRODUCTION

When asked to write this chapter, we were faced with two large and seemingly distinct tasks. The first was to review the emergence of information technology and its impact on curriculum and teaching in educational institutions, and in particular, on teaching and learning in science education. The second was to examine the roles played by evaluation in the development and assessment of sound principles which underlie the effective use of information technology.

In order to do justice to both we decided on an approach along the following lines. First, we examine how evaluation has impinged on curriculum and teaching at the school level up to now. The term school level is used advisedly because our view is that the use of information technology in science classes is strongly affected by the decisions taken by the school as a whole. Having set the school level scene, we then examine the roles

John M. Owen and Gerald T. Calnin, Centre for Program Evaluation, University of Melbourne, Parkville, Victoria 3010, Australia.   Faye C. Lambert, St. Michael's Grammar School, St. Kilda East 3163, Australia.

Evaluation of Science and Technology Education at the Dawn of a New Millennium, edited by James W. Altschuld and David D. Kumar, Kluwer Academic / Plenum Publishers, New York, 2002.

and impact of evaluation on school science education. These examinations are retrospective, with an emphasis on developments over the past decade. With these background conditions place, we are then in a position to move to a proactive stance. The conclusion sets out an agenda for evaluation of information technology in science education for the future.

While this chapter has a global focus, it has a distinctly Australian perspective. This is due to the fact that the authors are Australian, that some of the studies which we draw on were conducted "down under", and that we have actually undertaken evaluative studies of information technology in Australian schools. In addition, we have developed some innovative approaches to evaluation practice which we would like to share with an international audience. It is worth adding that a brief review of developments around the world suggests that educational technology issues facing Australian schools are very similar to those facing schools in the United States of America and many countries (Meredyth *et al.*, 1999). Thus while this chapter has a distinctly Australian flavour, there is clear evidence that the findings will be relevant to educators world wide.

EMPHASIS ON INFORMATION TECHNOLOGY IN THE CURRICULUM

Many developed and developing countries are placing an increased emphasis on information technology as a means of supporting the transition to a knowledge-based economy. Policy directions for schools is a manifestation of this trend; for example, *Getting America's Students Ready for the 21ˢᵗ Century* is a document designed to encourage the use of information technology in schools and to increase student achievement (USDE, 1996). Singapore has set out strategies for integrating information technology into education in four areas through its *Masterplan for Information Technology in Education* (SME, 1997). These areas are: curriculum and assessment, content and learning resources, physical and technical infrastructure, and human resource development. Other countries as far apart as Finland, Thailand and New Zealand have placed information technology on the national educational agenda.

An assessment of these policy developments shows their recency. Most if not all countries are only beginning to think through the implications for implementing the widespread use of computers and other related technologies in schools. Even in the richest country in the world, it has been found that

> few schools have adequate numbers of modern computers or access to the Internet, and relatively few teachers are prepared to use technology effectively. Further, access to computers and other technologies is not enough; integration of technology is also needed (USDE, 1997).

If this conclusion is extrapolated to the global level, one could infer that relatively little has been achieved so far regarding effective programs at the school level if an objective is that schools are to be a major influence in the quest for information technology literacy for citizens. In these circumstances, how can evaluation assist us in the promotion of effective policy and sound educational practice?

One way is to look for educational systems, schools and teachers and, for this chapter, science teachers, who are "ahead of the game", and to study them in detail. These are entities which have committed themselves to providing both material and human resources to information technology such that the achievement of desirable student outcomes is possible. Developments in Australia have provided circumstances that have encouraged such a scenario. Many private non-government schools in the country have committed themselves to extensive investment in computers and associated educational technologies, partly as a "marketing ploy" to attract the more affluent. As a reaction to this, government education departments have allocated extensive funds for technology in schools. To exemplify these trends, there are more than a dozen private schools located in Melbourne, Victoria, in which **all** students have access to a laptop computer for use in all classes. Although the Victorian Government cannot hope to match this level of provision in publicly funded schools, it had provided over 12,000 laptops to teachers by the end of 1999. Also, the ratio of students to computers is 5.7:1, the lowest figure across Australia, and we suspect, among the lowest ratios on this indicator world wide (Department of Education, Employment and Training, 1999).

EVALUATION AS CLARIFICATION

The impact of computers in education has been subject to a number of evaluations and studies. Evaluation theorists, however, generally agree that evaluation has a different meaning according to the purposes of a given evaluation exercise. While once seen to be associated with the determination of the worth of an intervention and in particular with the achievement of objectives, evaluation is now seen to perform a range of roles (Patton, 1996).

A key contribution evaluators can make during the formative or prototype stage of program development is to clarify the essential components of the program (Cronbach, 1980). Clarificative evaluation is concerned with:

- description of programs
- analysis of the logic or theory of programs

- plausibility of program design
- consistency between program design and implementation, and
- provision of a basis for subsequent program monitoring or impact evaluation. (For ease of discussion we will use the term program as an umbrella term for policy, programs and projects hereafter.)

Some of the key approaches to clarification evaluation include evaluability assessment and program logic development (Owen, 1999, pp. 190–219).

Clarifying the design of a program has several important outcomes pertinent to different evaluation audiences. One is to make program planners and deliverers aware of the key features of the program for which they are responsible. A second outcome is to provide decision makers, planners and administrators with evidence from which they may make a decision about expanding the program within an organisation. Thirdly, a clarifying evaluation can provide information to other interested parties outside the organisation about the benefits of adopting the program. A series of individual clarificative studies can be most helpful in the development of policy advice and informing policy-makers in a given system. The use of the findings of individual clarificative studies relies on an inductive type of knowledge production in which the findings of site specific evaluations can be synthesized with a view to providing a framework for systemic change.

When we began to work on the evaluation of information technology in schools about five years ago, there was a strong case for the use of clarificative evaluation. We report the findings from two studies here as the basis for extending the debate about effective information technology curricula in schools. Both examples involve the use of individual notebook computers (laptops) by students, and continuous access to these computers throughout the school day and after school. We now regard this as an essential platform, a necessary but not sufficient condition for effective information technology education in schools. Continuous availability of computers is consistent with what we are ultimately expecting from a curriculum in the knowledge generation as it will encourage the full integration of information technology into the everyday work of students and teachers in classrooms and other learning venues.

Both studies had similar scenarios. This involved the introduction of a major innovation into a school: laptop computers with an emphasis on their use in the middle years of schooling (years 5–10).

Even though involved schools had a general view of what the computers might achieve, a comprehensive implementation plan had not been developed prior to the adoption of what was a complex curriculum change.

One reason was due partly to the school administration needing to make a quick decision to adopt the computers without the time or resources to work through fully the implications of their introduction. This is a common feature found in the literature on innovation impact (see, for example, Rogers, 1995). Another reason was the newness and the complexity of what was being tried. Our view that it would have been impossible to fully specify the program in advance, is consistent with conclusions drawn by program design theorists. They have concluded that a comprehensive statement about the characteristics of an innovation can only be developed during the trials or first stages of its implementation (Smith, 1989).

FINDINGS FROM THE EVALUATIONS

The evaluation findings were used to consolidate the program in the schools which commissioned these evaluations. The purpose of reporting the findings is to delineate the essential characteristics of an integrated computer curriculum which we will call the "infotech curriculum".

In implementation terms, the infotech curriculum is more than just an alternative to computer education approaches that have been traditionally offered in schools. There is a move away from a situation where the teacher has the major control over the knowledge acquired by students. The infotech curriculum is a quadratic involving teacher, students, content, and notebook use. In an infotech curriculum, students have individual access to their own notebook computer which is integral to the day-to-day learning activities planned by the teacher.

The computer is used "off and on" throughout the day and at home when appropriate. In the cases studied, the computer was in use for between 30 and 50 percent of the school day. Students come to regard the computer almost as an extension of themselves and are asked to take responsibility for the computer throughout the school year. This involves not only ownership of the machine but also of the intellectual property they produce through their interaction with it. The computer becomes a personal assistant, a modifiable tool which can be used to communicate ideas and to expand the nature and place of learning.

Use of a personal computer in the ways just described leads to a curriculum which replaces more traditional arrangements between teacher and student. It impacts on the style of curriculum planning, classroom management strategies, student response and methods of assessment. The students have control or power over knowledge within frameworks set up by the teacher. An infotech curriculum is well served by the homeroom arrangements which characterize elementary (primary) and some middle school settings. In these arrangements, students and the teacher are located in one room for the majority of their classes. The room is a base where most the teaching and resources, such as the computer, can be readily accessed. This means that time taken for setting up and putting away the computer is minimized, leaving more of class time for productive use of the computer in teaching and learning situations. Students in these settings become responsible for using the computer, not only for learning, but also for its storage and safety.

The infotech curriculum is thus very different from introducing notebooks as an aid to students within individual subjects taught by different subject specialists. In this scenario, the tendency is for notebooks to be used to embellish traditional cognitive achievement depending on the software available in those subjects, rather than to encourage the qualitatively different learning experiences believed to occur where computers are made integral to the curriculum and used across a range of subjects.

HIERARCHIES OF INFORMATION TECHNOLOGY USE

From a review of the literature and the evaluations at these two schools we believe that an infotech curriculum involves students in a range of computer usage. This has been supported by a national study of information technology in Australian schools, Meredyth *et al.* (1999).

Basic use involves activities such as manipulating a mouse and a keyboard, turning a computer off and on, creating a document and printing it. It was found that two-thirds of Australian students had mastered these kinds of use, and that most of them had learned such skills at home. More advanced skills involved using a computer for creative writing, manipulating data bases and spreadsheets, sending email and creating a multimedia presentation (powerpoint). Advanced use was mastered by 60 percent of the students.

Table 1 indicates a trend (reading down the Table from basic and advanced use to more sophisticated computer usage). Seven different types of use are described and arranged in order of complexity. These are: Support, Link, Resource, Tutorial, Curriculum Adjunct, Curriculum Alternative, and Exploration and Control. As one moves down from one level to another, the student faces more challenging tasks and is required to be more independent in action. A further implication of the Table is to determine how schools can reorganize the more traditional curriculum to provide opportunities for students to have access to technology and learning experiences which lead to mastery of each mode in the hierarchy and the consequent acquisition of skills associated with each kind of use.

Table 1. A Hierarchy of Computer Use in an Infotech Curriculum

Mode	Description
Support	Used to enhance the presentation of work. • Word processing • Spelling and grammar checking • Desktop publishing • Multimedia/authoring Used to enhance the management of data. • Accumulating data on databases and spreadsheets • Mapping, graphing and graphics for presentation
Link	Used for communication between individuals. • Emailing organizations for data • Contacting people within and beyond the school • Desktop video conferencing
Resource	Used to access information and other resources. • Researching on the Internet • Accessing electronic databases
Tutorial	Used to facilitate learning of specific knowledge and skills via feedback provided at the individual pace of the learner. • Drill and practice exercises (eg key-boarding exercises) • Cloze exercises
Curriculum adjunct	Used in a specific subject to facilitate the teaching/learning of that subject. • Atlases • Graphing programs • Music programs • Topic based programs • Student journals and diaries • Languages other than English programs • Monitoring experimental data
Curriculum alternative	Used as an alternative approach to currently accepted teaching learning approaches. • Mathematica • Cabri • Robotics • Other simulations
Exploration and Control	Used to facilitate testing out solutions, decision-taking and problem-solving. This mode relies upon higher order skills. • Activities using Microworlds • Simulated environments using Project 2000

Table 2. Core Benefit Claims of an Infotech Curriculum

Acquisition of Student Skills

Claim 1   Enhances the acquisition of literacy, word processing and creative writing skills.

Claim 2   Enhances the acquisition of data manipulation/research skills; data-bases, spreadsheets, use of statistical packages, simulation and solution generators.

Claim 3   Enhances the acquisition of problem-solving skills.

Classroom Dynamics

Claim 5   Encourages cooperation between students in the classroom, in particular, the propensity of more able students to work with others of lesser ability.

Claim 6   Encourages student reflection; allows students to compare their performance with others on the same task.

Claim 7   Encourages students to stay on task and complete the task.

Claim 8   Encourages a classroom context that is less judgmental with less emphasis on right and wrong and more emphasis on what works.

Claim 9   Encourages a shift from reliance on verbal learning strategies to an integration of verbal and visual learning strategies.

Claim 10   Allows students to learn independently at their own pace and using their own learning materials.

Overall Teaching/Learning Environment

Claim 11   Leads to a more rewarding teaching/learning environment.

BENEFITS OF THE INFOTECH CURRICULUM: EXAMINING THE EVIDENCE

As part of one evaluation we used a variety of sources, including the available research literature, to assemble claims about the benefits or outcomes which would follow if there was full implementation of an infotech curriculum. While these claims only applied to this one evaluation, we present them here, in Table 2, as a contribution to current thinking about effective information technology outcomes.[1] These are discussed below.

UNIQUE BENEFITS OF THE INFOTECH CURRICULUM

While it is conceivable that many of the claimed benefits could result from use of computers in schools in more traditional modes, the infotech curriculum has inherent advantages which increase the likelihood that the learning benefits described in Table 2 will be fully realized. These advantages include high computer portability, relatively unrestricted access,

---

[1] A fuller version of this section of the paper appeared in Owen and Lambert (1996).

a sense of individual ownership and an emphasis on use across a range of subjects.

In this evaluation we also identified benefits provided by the infotech curriculum which had not been included in the up-front claims. In evaluation terms, these may be thought of as "unanticipated" or "unintended" outcomes. They are important because they are unique to the infotech curriculum and would be difficult, if not impossible, to achieve via the more traditional computer use strategies. It was found that the infotech curriculum strategy:

- encouraged flexibility in the physical organization of the classroom
- allowed computers to be incorporated across the curriculum regardless of location of class
- removed computer use from the exclusive domain of the computer specialist and required that it became an integral part of general classroom activity
- encouraged students to learn independently at their own pace, using their own learning materials
- increased the direct participation by females students, and
- enhanced the acquisition of keyboarding skills (Owen *et al.*, 1993).

DISCUSSION

Research on the use of information technology suggests that there are two views on its implementation in schools. One is that information technology is a support for the existing curriculum. Consistent with this view is that computers are a useful tool to complement existing ways of teaching. Such an approach would see computers placed at the back of a classroom or in a dedicated computer laboratory. By the nature of these arrangements, computer use is limited to some students in the case of the classroom or to allocated periods of time, in the case of the computer laboratory. An implication is that teacher professional development will help teachers upgrade their skills to use particular hardware and software in the classroom. This scenario is currently a familiar one. It implies a commitment to a traditional curriculum but in fact the limitations are set by resources available to students and teachers in the form of information technology availability.

The second view is more in line with the discussion of this chapter to date. This places information technology as an integral part of learning. Our evaluations of the infotech curriculum were designed to move the schools involved on the road to achieving a seachange in the way they conducted their teaching and learning. In these cases, the use of information technology is the key to achieve reformation of what happens in classrooms. Teacher professional development must encourage a fundamental rethink about things that have been held dear to many in the past, such as the division of the curriculum into subject areas, the traditional role of the teacher as a didactic teacher and the dominance of "broadcast learning" (Tapscott, 1998). Findings from the evaluation studies we undertook suggest that teachers need to be clear about what level of complexity they are invoking when working with students under this approach. We must be under no illusions that an infotech curriculum is a very complex intervention and in order to achieve widespread implementation will require unstinted support and encouragement from educational systems over a long period of time.

So where does this lead in the use of information technology in the science curriculum? And to be consistent with the thrust of this chapter, what have evaluation studies told us about the impact of information technology up to now? The next section examines these issues, and places the findings alongside those which we have outlined up to this point.

EVALUATING THE IMPACT OF TECHNOLOGY IN THE SCIENCE CURRICULUM

One of the most interesting features of educational reform in the last decade has been the almost universal view that computers and associated technology would change the nature of teaching and learning in school-based education in fundamentally powerful ways, more than any previous innovation since the introduction of mass education. Expectations have been grounded in social and economic developments where computers have played a central role in changing significantly the way we live and work. Despite these expectations, and the enormous investment of resources into the provision of computers in education, there is surprisingly little evidence that information technology has fundamentally altered the way in which teaching and learning occur in schools. There is some evidence, as indicated earlier, that an infotech curriculum is being introduced into a small number of schools, but the vast majority of approaches adopted are, in effect, a supplement to the traditional curriculum, and are based on traditional pedagogies and beliefs about student learning. We conclude that the computer continues to be viewed as having the potential, as yet unrealized, to change how we teach and learn.

Both the complexity of the innovation of a computer-integrated curriculum and the lack of a fully developed implementation strategy means that schools have devoted considerable time and energy to the resolution of issues associated with implementation. The demand for schools to negotiate these issues has meant that less emphasis has been given to the substantial issues such as student learning. A number of observers (Wiesenmayer & Koul, 1999; Papert, 1993; Sewell, 1990; Cuban, 1989) have commented on the regnant paradigm of the use of computers in schools: that they are still being used to supplement traditional classroom pedagogy and organisation, underpinned by a traditional epistemology, and that there is little evidence that schools are embracing the technology in ways which transcend the view of computer as tool or aid. At present there is little evidence of the transformational power of the computer in relation to student learning. Studies in Australia by Shears (1995), MLC (1993), Rowe (1993), Owen and Lambert (1996) have sought to illuminate and broaden our understanding of the impact of laptop computers on the curriculum and the learning process.

As indicated earlier, there are a range of ways in which computers can be used in an infotech curriculum. This also applies to the science curriculum and the science classroom. We identified seven categories or modes of use ranging from support in the classroom (such as, use for presentation of material or managing data) to exploration or control where the student is actively engaged with the computer and working towards the development of higher-order skills such as problem-solving. Clearly each of the modes has a place at different times and for different learning objectives in the science classroom; for example, database construction or graphing results can be achieved more efficiently with the use of particular software applications. So too, collecting data from a range of sources and communicating these to others can be achieved effectively through the use of information and communication technologies. What the evidence does suggest, however, is that the primary ways in which computers are being used in the science classrooms is in the support or supplementary mode; that is, as a tool to support and enhance traditional ways of teaching and learning.

For example, Campbell (2000) reports on a project where there was a conscious attempt to increase the use of communication and information technology in all science classes throughout the high school years. He concludes that there are a number of benefits to students and teachers, but that it is difficult to determine the cognitive gains for students. In Year 11 Physics, where computers were used to introduce spreadsheet simulation, Campbell found that it was necessary to "place these computer 'cognitive' activities into a context supported by relevant practical activities, which may be less efficient 'cognitively' but which have many 'values' to add [such as socialization, practical skills, linking to 'reality', and even enjoyment of the subject]" (p. 28). There is a strong belief that there are benefits for students, cognitively, but that they are difficult to identify as an independent variable in the learning equation, and that there is much to be gained from varying the learning environment as a stimulus to learning.

Similarly, Ainley et al. (1999) monitored a two-year program where students used laptop computers during the first two years of high school (years 7 and 8). In discussing their findings about use, the authors identify the primary modes employed by the students were those used to complete tasks (that is, word-processing etc.), those associated with learning how the machine worked (in essence, computer literacy), and, much less frequently, the ways in which "information and knowledge (were) being handled with the computer" (p. 65). The authors were equivocal about the benefits associated with the use of laptop computers in terms of improved student learning outcomes, though the evidence for improvements in the affective domain of motivation and enthusiasm for learning were strong. Jarvis et al. (1997) support this view and suggest that the use of computers in science classrooms has aided the learning of skills in the use of the computer, a form of computer literacy, rather than in the area of learning specific scientific processes.

So, though we are finding that computers, through productivity and efficiency, extend and supplement human performance and improve the affective domain of learning, a more profound implication would be that the computer can provide a tool which is capable of changing the characteristics of problems and learning tasks, and hence play an important task as a mediator of cognitive development.

Cognitive skills especially higher order thinking skills are those which we associate with synthesis and analysis, critical thinking and evaluation of ideas, hypothesizing and hypothesis testing, questioning techniques, observing patterns and underlying structures, making generalizations, and problem-solving strategies of planning, data collection etc. Indeed, though there are many definitions of critical thinking, it may be considered as "reasonable reflective thinking that is focused on deciding what to believe or do" (Ennis & Norris, 1990, p. 44). If we are to promote a learning environment which nurtures such skills, we need to create an environment where students actively participate in making decisions based on reflective thinking and evaluation of outcomes. When students become engaged in such processes they are not passive consumers of knowledge but active participants in their own learning. Therefore, their learning can occur at a deeper level of cognitive processing.

The promise of improving students' thinking skills has sparked considerable research. Much of the early research resulted from Papert's (1980) development of Logo, software designed to teach programming to students. Logo was claimed to enhance human cognition and the development and transfer of specific skills such as "planning skills" which are seen to be integral to problem solving. Again, as above, the evidence has, to date, been contradictory.

Perhaps the most promising research in this area suggests that programming accompanied by specific instruction on cognitive operations can be beneficial. Bangert-Drowns (1993) reports that "Logo programming was found to facilitate problem-solving and knowledge of geometry, respectively, when the programming was accompanied by instruction that capitalized on special strengths of Logo related to the instructional goals" (p. 71). Other research would suggest that another benefit of Logo is that it is not only a medium for skill learning but a medium for metacognitive thinking (Fletcher-Flinn & Suddendorf, 1996; Rowe, 1993) which is seen to transfer more easily than specific skills. Yet, concern has been expressed about the amount of time required for students to gain the level of programming skills which will enable them to acquire higher order problem-solving ability (Palumbo, 1990).

Other software are being developed to explicitly teach thinking and problem-solving skills. Ryser et al. (1995) point to the improvement in students' critical thinking ability when they are exposed to a Computer-Supported Intentional Learning Environment (CSILE). Abidin (1997) demonstrated that the problem-solving performance of students improved as a result of an Interactive Learning Environment (ILE); and Al-Anezi (1994) observed an improvement in thinking and classification skills using ILE. Geban et al. (1992) reported on the improvement in problem-solving skills through the use of simulation software.

Key difficulties in determining shifts in thinking and problem-solving are that these skills develop slowly and are difficult to measure. Another difficulty in teaching problem-solving skills is the degree to which they are able to be translated or transferred to other novel learning situations or to "real life" situations in science instruction. Simulation software has been shown to improve decision-making skills and, additionally, what was learned in a simulation usually transfers well to real situations (Alessi & Trollip, 1985). In this view Jacobson and Spiro (1995) have concluded that the use of hypertext systems has improved both memory of factual knowledge and promoted knowledge transfer enabling students to use the knowledge gained in new ways and in new situations.

Despite the powerful appeal of the claim that computers have the capacity to teach "higher order" thinking skills which will transfer to new learning situations, the evidence to date remains unclear, though there are promising signs (Shaw, 1998; Yerrick & Hoving, 1999). Research on the utilization of higher level skills like problem-solving suggests that in order to apply such skills it is necessary to have considerable specific knowledge about the content area in question. Further, several conditions seem to be required: "that learners possess the required domain-knowledge; that the situation to which transfer is expected is similar to one previously encountered; and that learners recognise that similarities exist" (Sewell, 1990, p. 213). This resonates with schema learning theory (Rumelhart, 1980) which affirms that, for knowledge to be constructed, students must be engaged in a rich learning environment where they are able to make connections between prior knowledge and the new input of information. It is the quality of the learning environment, the interaction between learner and technology or learner and teacher, which will best facilitate learning and the application or transfer of skills.

Simulation exercises, modeling and visualization (examples of Exploration and Control in the levels of computer use in Table 1) offer a number of potential benefits in the science classroom. In the first instance they provide access to important dimensions of a learning environment—they allow students to be actively involved in the learning process, allow students to learn and proceed at their own pace, and provide an opportunity to apply or integrate newly acquired knowledge. Importantly, in creating virtual experiments and problem-based microworlds, computer simulations allow students to "monitor experiments, test new models and improve their intuitive understanding of complex phenomena" (Akpan & Andre, 1999, p. 109). Computer-based simulations can also provide students with "problem situations that incorporate numerous variables, foreshortened time, and safe access to consequences of decisions and choices" (Okey, 1995, p. 84). Plus they have the advantages of access to areas which are impractical such as genetics experiments and a cost benefit edge in providing access through simulation rather that real-life alternatives.

Simulations and other forms of modeling or visualization offer students, when supplemented by effective teaching, opportunities to exercise the higher-order skills in reasoning and problem-solving educators had hoped that the new information technologies would realise. Barnea and Dori (1999) confirm in their research in high school chemistry classes that cognitive aspects of the "average student population" can be improved and that student learning "can be improved by using a discovery approach in a computerized learning environment" (p. 257). Similarly Sadler *et al.* (1999), deploying simulation software in a physics classroom, found that students had greater control over their learning, were able to model real-world systems, to make choices and test predictions about the behaviour of systems in a variety of configurations. The authors concluded that the skills of testing variables, making connections, and separating discreet behaviour patterns were all improved by the simulation software.

The infotech curriculum has the potential to incorporate student use of the Internet. Post-Zwicker *et al.* (1999) report that "student collaboration and student-led learning is encouraged when using the (internet) in the classroom" (p. 273), emphasizing the importance of active learning which occurs in a student-centered learning environment. Dorit and Fraser (1994) also point to the importance of an inquiry-oriented learning environment in their research involving computers in the science classroom. They found that in this environment "students perceived their classes as more investigative and open-ended, and their enquiry skills had improved" (p. 65).

Akpan and Andre (1999) investigated the importance of simulation exercises in the science classroom in relation to the sequencing of learning activities. They examined the prior use of simulation of frog dissection in improving students' actual dissection performance and learning of frog anatomy and morphology. They experimented with three conditions: simulation before dissection, dissection before simulation, or dissection only. They found that "students receiving SBD (simulation before dissection) performed significantly better" (p. 107) than students in the other categories on both actual dissection and knowledge of the anatomy and morphology. They state that "the effectiveness of simulations is dependent upon the sequence of presentation of learning activities to students" and that "presenting a genetic simulation before lecture enhanced learning more than the same simulation presented after lecture" (p. 117).

This research complements research completed by Akpan in 1998. He reviewed nearly 50 studies in the area of computer simulations in science education and noted that simulation can turn a tedious task into one that can be done more easily, cheaply or quickly, thus simulation can improve both efficiency and productivity in the science classroom. Akpan & Andre (1999), in examining other research, suggest that most research studies show that the use of "interactive video simulation as an instructional tool either improves learning or shows no difference when compared to the conventional method of instruction" (p. 110).

Other authors are much less sanguine about the benefits associated with the use of simulation software in the classroom. While there may be agreement about some of the benefits associated with the lower order skills (that is, data collection, construction of databases, graphing results etc.) and the associated benefits in terms of productivity and efficiency, there is little agreement about the capacity of the computer-simulation to develop higher-order skills of problem-solving and creative or lateral thinking.

Stratford (1997), in a detailed analysis of the research conducted during a ten year period on computer modelling and simulations to aid science instruction, observed that there "is no firm consensus yet as to all the benefits which might accrue to learners as a result of running simulations or constructing models. Neither is there agreement as to which instructional strategies might make the most effective use of computer-based models" (p. 19).

Beyond instruction, other research indicates that there are opportunities for computers to assist in performance assessment in science learning. While there are many areas in science assessment, from the simple recording of results to more complex tasks, performance assessment tends to focus on student achievement in problem-solving, higher-order thinking skills, and the application of skills and knowledge. Okey (1995), for example, argues that computer-based assessments can address the full range of outcomes to be learned in a science program. Baxter (1995) posits that one of the strengths of computer-based simulation used for assessment of student learning is that the computers can "maintain a full record of performance, so teachers and students can review problem-solving processes" (p. 21). In a simulated environment students are involved in a cyclical process of hypothesis testing and refining and the re-interpretation of observations: computers can track these processes and assist teachers in the analysis of problem-solving and thinking processes. Kumar and Helgeson (1995) detail the different types of computer use for science assessment and are optimistic that one of the promising areas for development is in the area of "solution pathway analysis" which permits the analysis of the steps a student takes in solving a problem. They note that computers offer an environment for "developing systems for assessing nonlinear problem-solving tasks. Based on cognitive science, a variety of rationalizations could be made with regard to the role of computer technology in such environments" (p. 33).

As described earlier, an infotech curriculum has become a reality for a small number of schools. We believe that this finding can also be applied to a small number of science classrooms. Nonetheless, the overwhelming evidence is that we have not yet made the progress expected a decade ago, despite the belief in the potential of computers to change, fundamentally, the nature of teaching and learning. Charp (1997) supports this view and underscores the idea that though there is a growing recognition of the importance of the use of technology in the classroom to aid learning, coupled with an increase in hardware and resources, "its integration into the curriculum is a slow process. Technology is still seen as an 'add-on' instead of an integral part of the curriculum." In a similar vein, Deal (1999) notes that, despite the potential for significant changes in the teaching and learning environment, "technology infusion into the curriculum may need new guidelines" for all science educators.

On the other hand, some authors remain optimistic that "the integration of technology into the curriculum (will provide) expanded learning opportunities . . . and give all students a learning environment that allows discovery and creativity . . . and has the potential to dramatically change the way we view science and mathematics" (Kimmel & Deek, 1995, p. 329; see also Berlin & White, 1995). For the computer to be a tool for the transformation of learning there will need to be considerable more input into teacher education and in-service training as well as resources for well-targeted research (Shaw, 1998). There is both optimism and pessimism as stated below:

> How far we seem to be from where we want to be in terms of student learning and attitudes . . . from the very experiences we report herein, the main body of data available to us is far less positive or encouraging than we would like it to be. As experienced teachers, as technology users, and as scientists who foresee drastic changes in the kind of intellectual skills our students are likely to be expected to bring to complex physics problems, we see a long development road ahead (Runge, 1999, p. 43).
>
> If one's goal is to promote inquiry through data collection and analysis of real world problems . . . or to establish discourse communities in which students and professionals construct knowledge . . . there are few superior tools than the evolving technology of the day (Yerrick & Hoving, 1999, p. 304).

EVALUATION: WHERE TO FROM HERE?

Central to educational reform is the notion that we need to equip students for life long learning, and that fundamental to this objective, is the development of thinking abilities. Few would debate that information technology skills are important for accessing information in a digital world and the promise of an infotech curriculum is that it offers much more than technical skills. For some time, the infotech curriculum has been heralded as a catalyst for change from didactic instruction and a content-driven curriculum to one with a focus on the individual student constructing his/her own knowledge in the process of inquiry—in short, a means to improving the quality of thinking skills, and even developing qualitatively different thinking skills (Norton & Resta, 1986).

Thinking abilities are at the heart of science teaching. An assumption which underlies education in science as a core study is that scientific literacy is important for all citizens:

to enable them to use scientific principles and processes in making personal decisions, to equip them to interact in a society which is increasingly dependent on scientific and technological skills and to give them genuine access to debate about scientific issues that affect society' (Woodleigh Science Curriculum, 1999).

Yet, there have been real concerns about the quality of science teaching and the extent to which these goals are realized. The basis for these concerns were confirmed some years ago in research cited by Adey and Shayer (1994). In a large study of 14,000 pupils in 45 English and Welsh schools, not more than 30% of 16 year olds were demonstrating even formal operational thinking (higher order thinking) which is "required to test hypotheses effectively, to judge critically between the merits of two arguments or to cope with proportionality" (McGuinness, 1999). The implication is that the majority of pupils could not think at a level required for the science curriculum, or indeed, for many other disciplines.

Educators have increasingly looked to information technology as a vehicle or tool for addressing the need to develop thinking skills, and evaluators have been asked to respond to two questions:

- Do students learn (and think) better as a result of computer-mediated learning experiences?
- Is the investment in technology "worth it", in terms of its impact on learning?

Earlier in this chapter we provided an overview of evaluations focused on the impact of computer-mediated learning in both science and more general educational contexts. From this overview, it is evident that evaluation studies have occurred within two different contexts:

- The first is the evaluation of an infotech curriculum in what might be described as "ordinary classrooms". Normal sized classrooms, (usually 25+ students) with teachers experienced in traditional curriculum but now working with students who have access to computers for their classroom work.
- The second is the evaluation of infotech curriculum under what might be described as "special circumstances". These circumstances generally include smaller class sizes, technical support and teachers identified for their strong understanding of pedagogy and how technology might be used to support that pedagogy.

EVALUATION OF INFOTECH CURRICULUM IN "ORDINARY" CLASSROOMS

This is typified by the Ainley (2000) study carried out in Victoria, Australia (referenced earlier in this chapter). A similar study was done by Beaufort County School District in the American state of South Carolina (1998).

Such studies tend to be short-term, the data collected are frequently based on surveys of students, parents and teachers leading to some description of attitudes and "use" based on self report. There is little or no observation of the "implementation" of the intervention, nor reference to the experience of the teachers involved and little attention is given to defining or measuring the types of outcomes that might flow from such an "intervention". Such studies have typically found that those involved in infotech environments believe that the technology makes a positive difference in the learning environment. This, Stevenson (1998) argues in the Beaufort County District evaluation, is important, but only a starting point:

> Perception, to some extent, is reality. If individuals believe something is making a difference, they tend to work consciously, and unconsciously in many cases to affirm their belief. Thus the fact that students, teachers, and parents believe in the benefits of the laptops better assures they will be used effectively. However, laptops are instructional tools. While "feeling" that computers are making a positive difference is a good starting point, policy makers, including school boards, also need hard data about the actual impact of this technological initiative.

Stevenson also attempted to measure impact on academic achievement with nationally standardized Metropolitan Achievement Tests, seventh edition [referred to as the MAT7]. Although there were indications that students performed better using laptops than without them, the relationship was subtle and complex and differences in achievement between the participating and non-participating students existed before the project began.

Such studies provide "indicators" that there is value in the use of technology in ordinary classrooms and some justification for continuing their use but are limited, both by their lack of attention to describing how the technology is integrated into the curriculum and to clarifying and measuring outcomes. They are indeed, only a starting point.

With these limitations in mind, we attempted to focus teacher perceptions more directly on learning outcomes in an evaluation of infotech curriculum in "ordinary" classrooms at Wesley College, a large private school in Victoria, Australia (Lambert et al., 1996). This example shows the complexity of the measurement challenges which will be faced by evaluators in the future in order to answer the question: Do students learn better using technology? Or more specifically, do students develop qualitatively superior thinking skills using technology?

The Wesley College infotech curriculum had been in place for three years and we were asked to look for evidence that students learn more effectively with the delivery of an infotech curriculum. Acknowledging the limitations of the methodology and a restricted time-frame, we developed a working definition of the outcomes which were designated as the focus for the study. One of these outcomes was "new ways of thinking/problem-solving". For each of the outcomes, we developed a series of "indicators" that might constitute evidence that improvement had occurred. This work was grounded in our reading of the literature on effective teaching and learning as well as the use of technology in the classroom and was informed by our direct observation of notebook programs at Wesley and other schools. Identifying the indicator for problem-solving required defining what was meant by problem-solving and then reaching some agreement as to what this activity might look like when it occurred in the context of a classroom.

We proposed that problem-solving should be regarded as a process in which a sequence of actions is developed to achieve some goal (Cohen & Feigenbaum, 1982). The sequence involves: defining the problem, planning what has to be done, gathering information, using "what-if reasoning", testing hypotheses, observing patterns and underlying structures and making generalizations. It frequently demands a high level of flexibility in thinking. Programming a computer is a form of problem-solving.

This reflects Ausubel's (1968) understanding of problem-solving, which he refers to as:

> any activity in which both the cognitive representation of prior experience and the components of a current problem situation are reorganised in order to achieve a designated objective. Such activity may consist of more or less trial-and-error variation of available alternatives or of a deliberate attempt to formulate a principle or discover a system of relations underlying the solution of a problem (insight).

The Ausubel definition underlines two types of problem-solving activity:

- the "making sense of" approach where the student goes through a process of trial and error which builds towards a goal of understanding the situation
- the "goal-directed" approach which involves a more formal process of devising and implementing a plan and usually involves the application of a previously learned rule or principle. The solution to the problem is then incorporated into the student's rule system as a higher order rule which can be reproduced and applied either to a different problem of the same type or related problems' (Evans, 1986, pp. 84–85).

In either mode of problem-solving, the computer offers another dimension to the acquisition of this skill. Evans argues that because the computer is such a flexible medium, any given problem often has a multitude of solutions and allows the exploration of different routes towards the solution. Without a computer, the exploration of alternatives is rarely feasible. In a print culture retracing one's steps requires a considerable amount of work. Hence, in a print culture, the motivation for exploring alternatives when a solution that works has been found is unlikely to be high because of the work involved.

Norton and Resta (1986) have suggested that the style of thinking involved in problem-solving in a print-oriented culture is qualitatively different from the style of thinking in a computer-oriented culture. The differences are:

- linear, sequential reasoning versus recognition of patterns and connections that result from relationships observed and tested
- deductive reasoning versus inductive reasoning
- "if-then" reasoning versus "what-if" reasoning
- reasoning in a closed system versus reasoning in an open system.

Based on this conceptual understanding of what "problem solving" was, we then developed indicators for use in this evaluation, which are summarised in Table 3 below.

Table 3. Problem-Solving: A Working Definition

Outcome	Key Indicators
Definition: Problem-solving is a process in which a sequence of actions is developed to achieve some goal. It involves: • defining the problem • planning what has to be done • using "what-if" reasoning • testing hypotheses • observing patterns and underlying structures • making generalizations • flexibility in thinking.	• Asking fewer how-do-I do it questions • Making use of trial and error discovery • Asking powerful questions of teachers and peers • Initiating inquiry • Showing comfort with ambiguity and open-ended assignments • Actively seeking the opinions of teachers and peers • Looking for underlying patterns and structures

This was a short-term outcomes study and is subject to the same limitations of the other attitudinal studies referred to earlier in that it was dependent on teacher perceptions. The purpose of its inclusion here is not to present this as a model, but to underline the need for future evaluation work to take on the challenge of measuring complex outcomes in ordinary classrooms. Any evaluation will need to explore more than levels of use or the degree to which skills have been acquired; evaluations will now need to grapple with the measurement of complex issues such as student understanding, problem-solving skills, or creativity.

EVALUATION OF INFOTECH CURRICULUM IN SPECIAL CIRCUMSTANCES

There are evaluation studies which have been carried out in situations where the groups of students are typically smaller than those in normal classrooms, where the focus for evaluation is often on specially designed software and where the programs are being implemented by teachers with a strong pedagogical base, working with researchers in specialized project designs. In such studies, the nature of the intervention is usually clear and there are explicit attempts to measure learning outcomes.

Some of these studies point to positive effects on the development of thinking skills, but within the findings in relation to cognitive gains remain unclear. What we have learned from these studies in special circumstances is that "computers alone do not mediate thinking and learning. Good software design and an appropriate pedagogy are crucial. Delivering information via computers is not sufficient for learning; knowledge (acquisition) needs to be designed" (McGuinness, 1999, p. 26).

Recognition of the critical role of teacher as "architect" of computer-mediated learning (Little, 2000) is notably absent from many of the evaluation studies that have been carried out in normal classrooms. If teachers must "architect" the learning, then it assumes that they have a strong pedagogical base. The lack of this base is particularly evident in the secondary school system where teachers are far more secure in their knowledge of the content of their own disciplines than in their understanding of student learning and how it might be facilitated. It is not surprising then, that the US National Science Foundation Teacher Enhancement Project identified the following principles which should guide the professional development of teachers using new technologies. They found that teachers need to:

- experience the process of inquiry in order to foster an inquiry approach in their classrooms
- learn new approaches to the content and teaching strategies in their subject areas as well as learning the old subject in new ways and to be able to reflect on this new learning
- be comfortable with the technology as a vehicle for inquiry
- be familiar with research on student learning
- work with other teachers on their professional development (KPMG, 1995, p. 21).

McGuinness (1999) argues that the real challenge is to transport the positive effects identified in small scale specialized projects into "ordinary" classrooms. The difficulties in accomplishing this should not be underestimated. Two important barriers to transfer relate to teachers and their understanding of their craft:

- teachers' craft knowledge can be threatened in their attempts to establish a constructivist learning environment
- teachers do not easily conceptualize the knowledge of their discipline in terms of students understanding and reasoning patterns (Leat, in McGuinness, 1999).

Other factors have also been identified as preventing the transfer from small scale to large scale projects. It has been found that special programs:

- only motivate students when they are special events and not part of everyday practices
- work well in small teacher/researcher to student ratios but not in classes with one teacher and 25+ students
- require unanticipated levels of professional development
- produce learning outcomes not valued within the context of competitive examination systems such as cooperative work, creativity and thinking skills (CTVG, 1996).

In the absence of specialized software, it is up to the classroom teacher to design learning experiences drawing on the interactivity that a computer environment provides to support and stimulate complex thinking processes. It is perhaps not surprising then, that the promise of infotech curriculum in enhancing thinking skills has not yet become a widespread feature in ordinary classrooms.

EVALUATION DESIGNS FOR THE FUTURE: TWO FUNDAMENTAL CHALLENGES

There are two major challenges which confront the evaluator in examining the role and impact of computer-mediated learning in science classrooms:

- the first relates to the complexity of use that surrounds the integration of computers into the classroom setting
- the second is the need to define and measure learning outcomes.

Complexity of Use

It is the complexity of use and context bound nature of computer use in classrooms that threaten the validity of simple input-output and control versus experimental designs. Crook (1994) warns us against conceptualising the computer in terms that suggest a medical model of how it works:

> Computers are unlikely to function as magic bullets—effortlessly releasing their therapeutic effects at points identified by teachers. The unfamiliarity and wizardry that surrounds them may cultivate such notions, but the real impact of learning through this technology may need to be measured with attention to how it is assimilated into the surrounding frame of educational activity (p. 9).

What Crook is underlining, is that the influence of the technology cannot be neatly contained within the events at the pupil-computer interface. The ways in which teachers interpret the relationship between the technology and traditional learning agenda is an important aspect of the impact of technology on learning settings as this then defines the context in which other changes might occur (Sheingold *et al.*, 1984).

In any study of an infotech curriculum in ordinary classrooms, it must be anticipated that there will be significant variations between teachers in the extent to which the technology has an impact on the restructuring of teaching and learning. Hence, evaluations in the future must involve *program clarification* as a means of illuminating how the learning experiences are structured—an assessment of outcomes must be linked to a rich understanding of implementation.

Outcomes Measurement

One of the most difficult aspects of evaluating the impact of technology on teaching and learning is defining and measuring learning outcomes. While there is general agreement that learning involves much more than information absorption or information processing (Wild *et al.*, 1994), significant questions and debate remain in regard to defining and measuring learning outcomes and isolating the impacts of the range of elements which influence learning.

Researchers working on the Apple Classrooms of Tomorrow Project (ACOT) were explicit about the challenges of evaluating an infotech curriculum. This longitudinal research, which was coordinated by UCLA and The Ohio State University, explored learning (traditional and non-traditional outcomes) when children and teachers had immediate access to interactive technologies both at home and at school. The research team initially applied conventional measures and short-term projects to assess student achievement but found that these were inadequate for assessing the complexity of the impact of use of technology on individual performance and achievement (ACOT Report, 1990). It is not surprising, then, that we found that *none* of the Australian schools that have led the world in making substantial investment in notebook programs where students all own their own personal computers, have also invested in systematic impact evaluation (Lambert *et al.*, 1996).

There remains the need for evaluations in the future to design appropriate measures to assess problem-solving skills, process writing and deep understanding. Such studies will need to draw on skilled and labor-intensive data collection procedures such as in-depth interviewing and observation in the context of longitudinal evaluation designs.

KEY EVALUATION QUESTIONS AND FORMS OF EVALUATION

We believe that evaluating the contribution of an infotech curriculum in the realm of science education will require two kinds of evaluation questions which demand distinctly different but complementary "forms of evaluation" work.[2] The first are encompassed by a form of evaluation described in the opening sections of this chapter—namely, *"evaluation for program clarification"*. Such questions might include:

- What does the implementation of an infotech science curriculum look like in the hands of highly information technology literate teachers with experience in its use in classrooms?

---

[2] The notion of Evaluation Forms is outlined in detail in Owen (1999). The argument is that there are five identifiable evaluation epistemologies which summarize current evaluation practice. As indicated earlier, two of these forms; (1) clarification, and (2) impact (particularly process-product studies) constitute the basis of this chapter.

- How is the infotech component incorporated into the learning context? That is, to what extent is it supported by, and/or linked to other methods of teaching?
- Is it likely that the "curriculum design in which infotech is incorporated" will lead to improved educational outcomes and/or qualitatively different educational outcomes, and more specifically, an improvement in complex thinking skills?

The second group of evaluation questions relate to the "*impact*" of the intervention, where we need to ask questions such as:

- Does an infotech curriculum lead to improved learning outcomes, and in particular, improvements in complex thinking skills?
- Does problem-solving in a computer environment lead to qualitatively different outcomes from problem-solving in a print culture?
- Does an infotech curriculum benefit all learners equally or does it have a different impact on different learners?

Both forms of evaluation are required in order to link outcomes with an understanding of implementation and both involve complex questions which require a range of expertise on the part of the evaluation team. Evaluation is often flawed in that it is often carried out by individuals confined by their specialist interests and skills. We advocate a team approach which is exemplified by a study currently being planned by Ainley at The University of Melbourne. This evaluation, which has a psychologist working with a learning specialist and an experienced classroom science teacher, examines a unit of work in the Physical Sciences and involves a micro-simulation which is nested within the context of other learning activities. The micro-simulation, designed by the teacher and students, has a range of parameters they control. The intention is to plot changes in learning and look for outcomes linked to the use of the simulation. Unit concepts are taught without the use of the technology, understandings are measured and then remeasured after the use of the micro-simulation.

CONCLUDING REMARKS

A fundamental difficulty with what is defined here as "ordinary" classroom studies is that the variable of teacher expertise is omitted from the equation. Computer technology will only lead to more effective learning, including higher order problem-solving when teachers have sufficient experience with the tool *combined with* insight into learning to build appropriate technology-rich learning experiences into the curriculum. The true impact of computers on learning is unlikely to occur if teachers try to simply overlay the technology onto existing curriculum practices. What is required—and what we suspect has yet to occur in any widespread way—is for teachers to rethink the learning experiences that they offer students and further, to explore how the technology might be used to support those learning experiences. Until this occurs, the introduction of technology into the curriculum as an innovation is not yet "fully implemented" and hence, as evaluators are all too aware, the validity of an outcomes study must be challenged.

Until we study what highly experienced, technologically literate science teachers can do with a class of students and access to technology, we will continue to be in a very poor position to make judgements about the worth of an infotech curriculum in science or elsewhere. This, we have argued, will require in-depth studies with experienced teachers by teams of researcher/evaluators who are prepared to tackle the difficult design questions posed by measuring learning outcomes.

It is evident that if evaluators are to make a contribution to the work of science educators and their use of technology to further science education objectives, they must identify and *describe* examples of good practice amongst those teachers who have been working in technology-rich environments for some time, those teachers *who are recognized* as being creative in their use of technology. But further than this, they need to acknowledge the important differences between implementation in the "ordinary" classroom and those which have special circumstances created for them. Studies in both are required.

It is also important that the findings of these studies become available to those who can influence change. The accumulation of evidence over time as a basis for action in the form of new policies and procedures has long been recognised. A synthesis of findings from studies described in this paper and in the previous paragraph should be useful to opinion leaders and change agents. One can think of university and college staff who have responsibilities for the professional development of science teachers as a key group in using such knowledge to advantage. Teacher educators, world wide, have a major role to play to persuade experienced teachers to revise their curricula and lobby for new teaching approaches in their schools. In addition they can and should influence beginning teachers and encourage them to use innovative teaching methods during practice teaching. There is a challenge here for science teacher educators around the world to think of ways of encouraging the use of an infotech curriculum, which takes into account school conditions as found in their country contexts.

REFERENCES

Abidin, B. (1997). Computer Based Representation for Aiding Problem Solving in Mathematics. Computer Based Learning Unit, University of Leeds.

ACOT Report 8 (1990). The Evolution of Teachers' Instructional Beliefs and Practices in High-Access-To-Technology Classrooms. First-Fourth Year Findings, Apple Classrooms of Tomorrow Advanced Technology Group, Apple Computer Inc, Cupertino, CA.

Adey, P., and Shayer, M. (1994). *Really Raising Standards: Cognitive Intervention and Academic Achievement*, Routledge, London, UK.

Ainley, M., Bourke, V., Chatfield, R., Hillman, K., and Watkins, I. (2000). *Laptops, Computers and Tools*, Australian Council for Educational Research Press, Camberwell, Australia.

Akpan, J. (1998). Computer simulations and learning science: In Akpan, J. and Andre, T. (1999). The effect of a prior dissection on middle school students' dissection performance and understanding of the anatomy and morphology of the frog. *Journal of Science Education and Technology* 8(2):107–119.

Akpan, J., and Andre, T. (1999). The effect of a prior dissection on middle school students' dissection performance and understanding of the anatomy and morphology of the frog. *Journal of Science Education and Technology* 8(2):107–119.

Al-Anezi, Y. (1994). Computer Based Learning Environments for Mathematical Classification Skills. PhD thesis: Leeds University.

Alessi, S. M., and Trollip, S. R. (1985). *Computer-Based Instruction: Methods and Development*, Prentice-Hall, Englewood Cliffs, NJ.

Ausubel, D. (1968). *Educational Psychology—A Cognitive View*. Holt Rinehart and Winston, New York, NY.

Barnea, N., and Dori, Y. (1999). High school chemistry students' performance and gender differences in a computerized molecular modeling learning environment. *Journal of Science Education and Technology* 8(4):257–264.

Bangert-Drowns, R. L. (1993). The word processor as an instructional tool: a meta-analysis of word processing in writing instruction. *Review of Educational Research* 63(1):69–93.

Baxter, G. P. (1995). Using computer simulations to assess hands-on science learning. *Journal of Science Education and Technology* 4(1):21–28.

Berlin, D. F., and White, A. L. (1995). Using technology in assessing integrated science and mathematics learning. *Journal of Science Education and Technology* 4(1):47–60.

Campbell, A. (2000). Implementing use of information and communication in the teaching of science in a small rural school. In Fawns, R., Rodrigues, S., and Sadler, J. (Eds.), *Science Research in Schools Programme 1998–2000*. Final Reports for the Nine Projects, The University of Melbourne, Department of Science and Mathematics Education.

Charp, S. (1997). Changing teaching strategies through technology. *Technological Horizons in Education* 24(10):6–8. *http:www.thejournal.com.magazine/vault/A1766.cfm*

Cohen, P., and Geigenbaum, E. (1982). *Handbook of Artificial Intelligence*, Vol. 3, W. Kaufmann, Los Altos, CA.

Crook, C. (1994). *Computers and the Collaborative Experience of Learning*, Routledge, London, UK.

Cronbach, L. J. (1980). *Toward Reform of Program Evaluation*, Jossey-Bass, San Francisco, CA.

CTVG (1996). Looking at technology in context: a framework for understanding technology and education research. In Berliner, D. C. and Calfee, R. C. (Eds.), *Handbook of Educational Psychology*, New York, Macmillan, pp. 807–840.

Cuban, L. (1989). Neoprogressive visions and organisational realities. *Harvard Educational Review* 59:217–222.

Deal, S. (1999). The cyber quest: a tool to assess educational resources on the internet. *Technological Horizons in Education* 26(10):50–57.

Department of Education, Employment and Training. (1999). Annual Report. http://eee.eduvic.gov.au/annual.htm. Government of Victoria, Melbourne, Australia.

Ennis, R. H., and Norris, S. P. (1990). *Evaluating Critical Thinking*, Hawker Brownlow, Cheltenham, Vic.

Evans, N. (1986). *The Future of the Microcomputer in Schools*, Macmillan Education Ltd, London, UK.

Fletcher-Flinn, C., and Suddendorf, T. (1996). Do computers affect "the mind"? *Journal of Educational Computing Research* 15(2):97–112.

Geban, O., Askar, P., and Ozkan, I. (1992). Effects of computer simulations and problem-solving approaches on high school students. *Journal of Educational Research* 89(1):1–10.

Jacobson, M. J., and Spiro, R. J. (1995). Hypertext learning environments, cognitive flexibility, and the transfer of complex knowledge: an empirical investigation. *Journal of Educational Computing Research* 12(4):301–333.

Jarvis, T., Hargreaves, L., and Comber, C. (1997). An evaluation of the role of email in promoting science investigation skills in primary rural schools in England. *Research in Science Education* 27(1):223–236.

Kimmel, H., and Deck, F. (1995). Instructional technology: A tool or a panacea? *Journal of Science Education and Technology* 4(4):327–332.

KPMG, (1995). Information technology in education and the arts. Unpublished report for the Department of Education and the Arts, Government of Victoria, Australia.

Kumar, D. D., and Helgeson, S. L. (1995). Trends in computer applications in science assessment. *Journal of Science Education and Technology* 4(1):29–36.

Lambert, F. C., Owen, J. M., and Preiss, A. (1996). The future of the notebook initiative, Unpublished report for Wesley College, Melbourne, Australia.

Little, J. (2000). Teacher as architect?—how to describe what we do in technology-rich learning environments, Unpublished paper, Hard Fun Technology & Learning Consultancy. Private Communication.

Maor, D., and Fraser, B. (1994). An evaluation of an inquiry-based computer-assisted learning environment. *Australian Science Teachers Journal* 40(4):65–70.

McGuinness, D. (1999). From thinking skills to thinking classrooms: a review and evaluation of approaches for developing pupils' thinking, Report for the United Kingdom Department for Education and Employment, London, UK.

Meredyth, D., Russell, N., Blackwood, L., Thomas, J., and Wise, P. (1999). Real Time. *Computers, Change and Schooling*. Australian Key Centre for Cultural and Media Policy. Griffith University and Queensland University of Technology. Department of Education, Training and Youth Affairs, Canberra. ISBN 0 642 23913 4.

Methodist Ladies College. (1993). Reflections of a Learning Community. Views on the Introduction of Laptops at MLC. Melbourne, Australia.

Norton, P., and Resta, V. (1986). Investigating the impact of computer instruction on elementary students' reading achievement. *Educational Technology* March 1986:35–40.

Okey, J. R. (1995). Performance assessment and science learning: rationale for computers. *Journal of Science Education and Technology* 4(1):81–94.

Office of Technology Assessment (OTA). (1995). Teachers and Technology: Making the Connection. Office of Technology and Assessment Press: Washington, DC.

Owen, J. M. (1999). *Program Evaluation: Forms and Approaches*. Second edition, Allen and Unwin and Sage, Sydney and London (with P. Rogers).

Owen, J. M., and Lambert, F. C. (1996). The notebook curriculum: an innovative approach to the use of personal computers in the classroom. *Australian Educational Computing* 11(1):26–32.

Owen, J. M., Lambert, F. C., and Hurworth, R. E. (1993). Notebook computers in the curriculum. Centre for Program Evaluation, The University of Melbourne, Melbourne.

Palumbo, D. B. (1990). Programming language problem-solving research: a review of relevant issues. *Review of Educational Research* 60(1):65–89.

Papert, S. (1980). *Mindstorms: Children, Computers and Powerful Ideas*, Basic Books, New York, NY.

Papert, S. (1993). *The Children's Machine: Rethinking Schools in the Age of the Computer*, Basic Books, New York, NY.

Patton, M. Q. (1996). A world larger than formative and summative. *Evaluation Practice* 17(2):131–144.

Post-Zwicker, A., Davis, W., Grip, R., McKay, M., Pfaff, R., and Stotler, D. (1999). Teaching contemporary physics using real-time data obtained via the world-wide web. *Journal of Science Education and Technology* 8(4):273–281.

Radish, E. F., Saul, J. M., and Steinberg, R. N. (1997). On the effectiveness of active-engagement microcomputer-based laboratories. *American Journal of Physics* 65(1):45–54.

Rogers, E. (1995). *Diffusion of Innovations*. Fourth Edition, The Free Press, New York, NY.

Rowe, H. (1993). *Learning with Personal Computers. Issues, Observations and Perspectives*, Australian Council for Educational Research, Camberwell, Australia.

Rumelhart, D. (1980). Schemata: the building blocks of cognition. In Spiro, R., Bruce, B., and Brewer, W. (Eds.), *Theoretical Issues in Reading Comprehension: Perspectives from Cognitive Psychology, Linguistics, Artificial Intelligence and Education*. Lawrence Erlbaum, Hilldale, NJ.

Runge, A., Spiegel, A., Pytlik, L., Dunbar, S., Fuller, R., Sowell, G., and Brooks, D. (1999). Hands-on computers in classrooms: the skeptics are still waiting. *Journal of Science Education and Technology* 8(1):33–39.

Ryser, G. R., Beeler, J. E., and McKenzie, C. M. (1995). Effects of a computer-supported intentional learning environment (CSILE) on students' self-concept, self-regulatory behaviour, and critical thinking ability. *Journal of Educational Computing Research* 13(4):375–385.

Sadler, P. M., Whitney, C. A., Shore, L., and Deutsch, F. (1999). Visualization and representation of physical systems: wavemaker as an aid to conceptualizing wave phenomena. *Journal of Science Education and Technology* 8(3):197–209.

Sewell, D. F. (1990). *New Tools for New Minds*, Harvester Wheatsheaf, London, UK.

Shears, L. (Ed.) (1995). *Computers and Schools*, Australian Council for Educational Research, Camberwell, Australia.

Shaw, D. E. (1998). Report to the President on the use of technology to strengthen K-12 education in the United States: findings related to research and evaluation. *Journal of Science Education and Technology* 7(2):115–126.

Soska, M. (1994). An introduction to educational technology. *Directions in Language and Education* 1(1):1–7.

SME (1997). Masterplan of IT in Education. Ministry of Education, Singapore.

Smith, M. F. (1989). *Evaluability Assessment: A Practical Approach*, Kluwer, Boston, MA.

Stratford, S. J. (1997). A review of computer-based model research in pre-college science classrooms. *Journal of Computers in Mathematics and Science Teaching* 16(1):3–23.

Stevenson, K. R. (1998). Evaluation Report—Year 2 Schoolbook Laptop Project, Beaufort County School District, Beaufort, SC.

Tapscott, D. (1998). *Growing Up Digital: The Rise of the Net Generation*, McGraw Hill, New York, NY.

USDE (1996). Getting America's Students Ready for the 21st Century: Meeting the Technology Literacy Challenge, United States of America, Department of Education, Washington, DC.

USDE (1997). Education 1997–1999: Government Strategy, United States of America, Department of Education, Washington, DC.

Wiesenmayer, R., and Koul, R. (1999). Level of internet use among science teachers involved in a professional development project. *Journal of Science Education and Technology* 8(2):123–135.

Wild, M., and Kirkpatrick, D. (Eds.) (1994). Computer Education: New Perspectives, Mathematics, Science and Technology Education Centre, Edith Cowan University, Perth, Western Australia.

Woodleigh School (1999). Woodleigh science curriculum, unpublished document produced for the School Registration Board. Woodleigh School, Baxter, Australia.

Yerrick, R., and Hoving, T. (1999). Obstacles confronting technology initiatives as seen through the experience of science teachers: a comparative study of science teachers' beliefs, planning, and practice. *Journal of Science Education and Technology* 8(4):291–307.

CHAPTER 7

Complementary Approaches to Evaluating Technology in Science Teacher Education

David D. Kumar and James W. Altschuld

INTRODUCTION

Science teacher education in the United States, as well as in other nations (e.g., India, United Kingdom) is undergoing major change. In the United States, strategies for increasing the pool of qualified teachers are being explored by teacher preparation programs. One possibility for providing better training for teachers of science would be to modify pre-service programs with special emphasis being given to the use of technology. Technology can augment student learning and can be employed for assessment purposes.

In regard to technology, there has been dramatic growth not only in its availability but also in its use and the use of related products designed

David D. Kumar, College of Education, Florida Atlantic University, 2912 College Avenue, Davie, FL 33314.   James W. Altschuld, Educational Policy and Leadership, The Ohio State University, 29 W. Woodruff Avenue, Columbus, OH 43210.

Evaluation of Science and Technology Education at the Dawn of a New Millennium, edited by James W. Altschuld and David D. Kumar, Kluwer Academic / Plenum Publishers, New York, 2002.

for the classroom and/or the preparation of teachers. Interactive video (IVD) technology, as an example, has been for some time, a focus of attention in general education as well as science education (Berger, Lu, Belzer, & Voss, 1994).

More than a decade ago, Pollack (1989) estimated the number of instructional videodiscs to be "560+", with undoubtedly many more such products being accessible, being produced or coming onto the market. Kumar, Helgeson and Fulton (1994) observed that about 29% of the teacher education programs in Ohio reported using interactive video technology in pre-service teacher education in science, and 92% of those pre-service programs incorporated it into the teaching of instructional strategies. Goldman and Barron (1990) and Vitale and Romance (1992) found that the use of IVD technology in pre-service teacher preparation boosted teachers' "confidence" in teaching, understanding of science knowledge, and attitudes toward science teaching.

Abell, Cennamo, Anderson, and Bryan (1996) reported the development of interactive video applications for reflective learning in elementary science methods courses. According to Abell, *et al.* (1996), "video allows pre-service teachers to enter a classroom virtual world and witness events as they occur" (p. 138). While the merging of technology and science teacher education would seem to be promising, there is no guarantee that it will lead directly to improvement in student enthusiasm and learning, and ultimately to modification in the classroom practices of elementary and secondary science teachers. A national survey in the United States commissioned by the Milken Exchange on Educational Technology (1999) and conducted by the International Society for Technology in Education found that most teacher preparation programs are not successful in incorporating technology into their curricula.

What might account for this observation? Perhaps, teacher educators are forced to accept technology on face value. Lack of in-depth evaluation about the quality of technology and the context into which it was incorporated could explain, at least in part, why teacher education programs experience difficulty in making decisions to adapt technology-based instruction for their situations. In support of this point, Grandgenett, Ziebarth, Koneck, Farnham, McQuillan, and Larson (1992) noted a paucity of evidence of the effectiveness of interactive videodiscs in pre-service teacher education, as demonstrated through carefully conducted evaluation studies.

Evaluations of technological innovations in science teacher education could provide insight, into the role of technology in science education and even more importantly, information necessary to shape policy decisions concerning building it into the pre-service curriculum. Put simply, science educators, technology educators and policy makers need more sophisticated understandings of educational technology as it pertains to pre-service science teacher education.

In this vein, the following discussion looks at an interesting and relevant case in which two distinct evaluations were conducted of a pre-service science teacher education project that had a heavy technology emphasis. Comparisons of the two evaluations reveal the complexity of evaluating such projects and the multifaceted ways in which the evaluation endeavor could be approached. It should be noted that having two unique assessments of the same project from totally different perspectives (a more traditional one as well as one that is somewhat non-traditional) affords an unusual situation and a seldom seen opportunity for thinking about evaluations. Additionally, a brief overview of what some other authors are telling us about the evaluations of technology in the teacher training is presented.

DESCRIPTION OF THE PROJECT EVALUATED

The project "Improving Science Education A Collaborative Approach to the Preparation of Elementary School Teachers," developed by and implemented at Vanderbilt University was an interactive media teacher education project funded by the National Science Foundation (NSF). It was aimed at redesigning the science (and mathematics) methods courses taught at Vanderbilt with the aid of interactive video technology. In project materials, real world practices of effective science teaching were presented to bridge the gap between science content and teacher education courses. Project staff conceptualized and produced videodiscs consisting of video-based lesson examples as a supplement to regular class lectures and assignments. One intent was to help students learn instructional skills and strategies that were identified as effective methods of teaching science. The project capitalized on the skills and experiences of science and math educators from the Peabody College of Education, science faculty from the College of Arts and Sciences, and consultant teachers from local area schools (grades 4–7).

Videotapes of classroom teaching episodes contrasting experienced and beginning teachers were systemically designed and incorporated into the videodiscs and constituted the core of the content emphasized by the project. A thirty-minute videodisc permitted elementary science methods students random access to video segments of classroom practices linked to a programmed Hypercard stack on a computer. "A card showing on the computer screen for example, might contain a paragraph of text describing a common misconception in science and a small picture of a camera (video icon) that allows the teacher or student to select a videodisc segment of a classroom scene depicting the misconception. The user moves a pointer around the screen by rolling the 'mouse' and selects an item such as the video icon by pressing the button on the mouse (Goldman & Barron, 1990, p. 24)." Over time, advanced versions of the videodiscs were produced that allowed students to view two or three ways of teaching the same concepts side by side on one screen at the same time. The ability to see different approaches like this is quite unique when looked at from what normally occurs in pre-service education.

When a student teacher is in the observation part of their training they only can see one classroom and one teacher with probably one general teaching style. Using the videodiscs, their perceptions and preconceived mind sets are seriously challenged and the possibilities for instruction begin to multiply in a startling fashion. The potential for changing pre-service instruction and for pushing student reflection on how to teach and how to vary what happens in school classrooms is enormous.

A TRADITIONAL EVALUATION OF THE VANDERBILT PROJECT

The Vanderbilt University interactive video project had been thoroughly evaluated by the project staff via what might be termed—traditional evaluation techniques. (See the report by Barron, Joesten, Goldman, Hofwolt, Bibring, Holladay, and Sherwood, 1993 for specific details of the evaluation. Only summarized highlights will be given here.) The sample for the evaluation consisted of students enrolled in the science methods courses with an associated practicum component and those in the elementary student teaching experience usually one or two semesters after the methods course. Students in the science methods course that experienced the interactive videodiscs formed the video group and individuals enrolled in the science methods course prior to implementing the videos constituted the baseline group. Data collection included pre- and post-tests, ratings of classroom observations of the teaching practicum and student teaching, and interviews of student teachers after they were observed. The observation ratings focused on Teaching Competency, and Student Behavior. Additionally, the percentage of class time devoted to particular activities and course examination scores were taken into consideration. Data were analyzed using descriptive and inferential statistics.

In Table 1, selected significant ($p < .05$) findings are listed from this evaluation as reported by Baron $et\ al.$ (1993). The statistical outcomes are t-tests based upon comparisons of pretest and post test outcomes for the groups specified in the Table. The findings could be classified into teaching competencies, student behavior, and activities. It is interesting that out

Table 1. Selected Significant (p < .05) Findings of the Traditional Evaluation of
the Vanderbilt Project reported in Barron, L. C., Joesten, M. D., Goldman,
E. S., Hofwolt, C. A., Bibring, J. B., Holladay, W. G., & Sherwood, R. D. (1993). *Improving
science education: A collaborative approach to the preparation of elementary school teachers.*
A final report to the National Science Foundation under grant number TPE-8950310.
Nashville, TN: Vanderbilt University

Variable Significant (p < 0.05) Means Based on Percent Classtime	Group with Observation Rating
TEACHING COMPETENCY	
1. Selecting materials/learning experiences which stimulate student curiosity and support their investigation	ST-V
2. Preparing for alternative ideas and situations	ST-B
3. Sufficient variety of manipulatives and materials to enhance understanding	ST-V
4. Questioning to clarify understanding	PS-B
5. Activities/lesson formats appropriate to the level of the learner	ST-V
6. Appropriate sequencing of content and pedagogy	ST-V
7. Monitoring understanding	PS-B
8. Clearly defining tasks	ST-B
STUDENT BEHAVIOR	
9. Student involvement in lesson	ST-V
10. Student understanding of purpose of instruction	ST-V
11. Students' interest in lesson	ST-V
ACTIVITIES	
12. Student activities (Discovery or Inquiry)	ST-V, PS-V
13. Individual seatwork	ST-B, PS-B
14. Procedural/Behavioral presentations	ST-V, PS-V
15. Transitions	ST-B

NOTE: ST-V = Student Teacher Video Group; ST-B = Student Teacher Baseline Group;
PS-V = Practicum Student Video Group; PS-B = Practicum Student Baseline Group.

of the eight significant teaching competencies, four involved the student
teacher-video group (e.g., "Sufficient variety of manipulative and materials
to enhance understanding") with the rest coming from either the student
teacher-baseline group (e.g., "Preparing for alternative ideas and situations") or the practicum student-baseline group (e.g., "Questioning to
clarify understanding"). The three significant student behavior variables
were produced by the student teacher-video group (e.g., "Student involvement in lesson"). On the other hand, the four significant variables
involving activities came from a combination of the student teacher-video

and practicum student-video groups (e.g., "Student activities discovery/ inquiry"), and baseline groups (e.g., "Individual seat work").

The evaluation results generated by Baron and her associates (1993) also contained a summary of responses to post-observation interview questions. Over 50% student teachers and practicum students have identified "Hands-on Instruction" and "Discovery" as the strength of their teaching experience.

A CONTEXT (NON-TRADITIONAL) EVALUATION OF THE PROJECT

This evaluation dealt with the Vanderbilt project according to the structure of a program evaluation model published by Altschuld and Kumar (1995). The history behind the model is predicated on a review of the science education literature particularly that part of it focusing on evaluation in the field. Altschuld and Kumar identified and examined sources particularly with regard to models that had specific features or adaptations related to the evaluation of science education programs. Few sources were obtained from the ERIC database over a 20 year time frame. From what they found and from the more general evaluation literature, Altschuld and Kumar developed the model depicted in Figure 1. In that framework, it is recognized that contextual and environmental factors would be important for the successful implementation, and indeed, the ultimate success of educational programs in general and science education programs in specific. See Exline

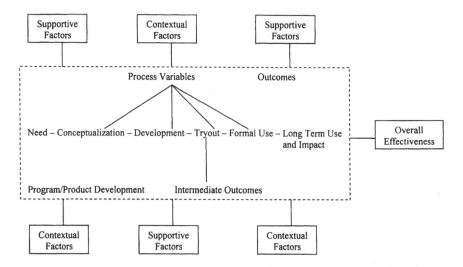

Figure 1. General Synthesized Evaluation Model Structure (Altschuld, & Kumar, 1995).

(1985), Exline and Tonelson (1987), Welch (1978), Stufflebeam (1968), Small (1988), and Field and Hill (1988) for additional discussion of the idea of environmental and contextual factors. The heavy stress on contextual and supportive factors is the key distinguishing feature of the model. It should be noted that after the completion of the initial work on the model, it was later augmented to include reinvention (Altschuld, Kumar, Smith, & Goodway, 1999).

In Figure 1, the development or life cycle of a product/innovation is the central core of what is to be evaluated. The model takes into consideration that every part of the life cycle is embedded within a complex milieu of social, economic, and organizational factors. Science education products and programs are heavily dependent on the support of these factors and they are often relied on to bring science to "life" (visitations to science centers or local science related businesses, invitations to scientists to come to classrooms, etc.) If such support isn't there, the likelihood of success will be significantly diminished during all phases of the life cycle. The focus derived from the model guided the second evaluation of the Vanderbilt interactive video project. (See Kumar and Altschuld, 1999) for an extensive and very detailed description of the non-tradition evaluation.)

An evaluation following the premises of the model would stress organizational climate, evidence that the environment was supportive of change and experimentation, the provision of resources to enable users to adopt and/or adapt new products and processes, administrative interest in and knowledge of the innovation or program, and other similar concerns. If these concerns are compared to the previously described evaluation of the project, it should be apparent that they are somewhat orthogonal to what was looked at in a more traditional approach. It isn't that one evaluation strategy is better than the other but that the two are being conducted from appreciably different world views or alternatively the "gestalt" is different and as such the evaluations will differ in noticeable ways.

Returning to the nature of the model, the measurement of variables is more amenable to qualitative methods, characterized by on-site visits with in-depth interviews and observations, rather than quantitative ones. Furthermore, an important feature of the evaluation conducted by Kumar and Altschuld was not to just explore the context and supportive environment of the Vanderbilt project but to do so long after its products had been developed and implemented. This follow-up type of evaluation, while critical to long term use, is generally not part of most evaluations. Since the project ended approximately two years before this context sensitive evaluation, it carefully investigated the press for continued implementation and the interaction of the project and its environment.

METHODOLOGY

The evaluation involved document review, on-site observations, development of interview protocols, sample selection, interviews, and analysis of interview data. After gleaning as much as they could from available materials about project enthusiasm, physical facilities, financial resources, principles imparted through the project, availability of other resources, and fit of the program into the environment, the following general questions were explored. What factors worked for success? What factors impeded and/or propelled success? What is happening in long-term use? How has the project evolved past the funding stage? How have schools adopted or changed to facilitate the project? What are the key environmental features affecting the adaptation process? From these categories, related but different sets of questions were developed for students, faculty, and project staff members.

Protocols employed in the first wave of interviews were revised to probe more deeply into some areas during a second site visit with an entirely new sample of interviewees. Thus, it was possible to examine and explore emerging hypotheses and understandings that developed from analyzing the data from the interviews conducted during the first visit.

The interviews included introductory comments explaining the rationale of the evaluation, the purposes of the interviews, and that all interviews were confidential with only grouped results being reported. Respondents were strongly encouraged to be open and frank in their comments. A few examples of the questions used in the interview protocols were:

1. "Are you personally involved? Explain how you and others became involved in this project?";
2. "In your judgment, what are or would be the long-term outcomes of the project and the use of interactive media; on students, on the faculty/academic environment?";
3. "If another teacher education institution, or another area were to adopt a project like this, what specifically would you recommend that they do or consider, and why?";
4. "What did you feel that you learned from the interactive media portion of the program? Please be as specific as you can in your answer."; and
5. "If you were telling other students about the integrated media aspect of the program, what would you emphasize in your discussion?"

With the assistance of the project staff at Vanderbilt, a sample of interviewees was identified and scheduled for interviews which took place during the two site visits. On average, each interview with students and former students (graduates) lasted about twenty-five to thirty minutes. Interviews with project staff members, directors, administrators, and faculty were about one hour in length. The two evaluators worked as a team with one leading the interview process. Short notes were taken during the interviews which were also audio taped for subsequent transcription. The evaluation team members continually verified the transcripts of the audio recordings of the interviews and the handwritten notes by being in frequent contact with each other as they independently analyzed (a form of an independent reliability check) the interview results. This check was a means of ensuring the accuracy of the data.

The analytic steps included coding, using the constant-comparative method for developing initial data categories, describing the categories, looking across them for emergent explanatory themes, and seeking support/substantiation for the explanatory themes. Due to the fact that qualitative studies generate reams of data, only samples of findings are provided in the next section. Hopefully they will serve to show the richness of results produced by this evaluation.

RESULTS (INITIAL DATA CATEGORIES)

The findings of the evaluations as classified into initial data categories and emerging explanatory themes are provided in Table 2. To give some feel for the data, selected initial data categories derived from one of the interviewed groups—Faculty, Staff, and Administrators—are as follows.

Student Learning. A greater conceptual understanding of teaching and learning, and making science teaching alive in a clear and concrete way was evident. The gains in understanding were attributed to hands-on methods in activities or planning for same as taught in the interactive, media based methods courses. Students were willing to tackle hands-on science lessons in their practicum paving the way for future use of hands-on science in classrooms.

Effects Upon Instructional Staff. Faculty members learned how to develop these types of materials and to examine and redefine the purpose and nature of elementary methods courses (over a period of time). The project helped to rejuvenate some faculty and to refocus their teaching. To some degree, the professional status of individual faculty members (presentations, and being asked for help, etc.) was enhanced.

Institutional Effects. Institutional benefits relate to the spread of the interactive approach to the biology course and to greater integration of

Table 2. Emergent Explanatory Themes and Initial Data Categories of the Non-Traditional Evaluation of the Vanderbilt Project reported in Kumar, D. D., and Altschuld, J. W. (1999). Evaluation of interactive media in science education. *Journal of Science Education and Technology*, *8*(1), 55–65

EMERGENT EXPLANATORY THEMES

1. Administrative Support
2. Osmosis/Permeation
3. Student Perception on Campus
4. Student Perception in Schools
5. Technical Support
6. Cross Fertilization
7. Organizational Climate
8. Critical Mass

INITIAL DATA CATEGORIES

I. Faculty, Staff and Administrator Interviews
   a. Student Learning
   b. Student Opportunities to Learn
   c. Instruction
   d. Effects Upon Instructional Staff
   e. Institutional Effects
   f. Major Thrust of Interactive Video
   g. Role of Interactive Video in Lab Experiences
   h. Vanderbilt University Administrator Support
   i. Use by Other Faculty
   j. Measurable Student Outcomes
   k. Technical Support
   l. Transferability to Other Colleges and Universities
   m. Transferability to Schools Employing Vanderbilt Graduates
   n. Difficulties

II. School Administrator Interview
   a. Supportive Environment in School
   b. Transfer Experiences from the University to the Classroom
   c. Support from Existing Teachers
   d. Probing, Reflecting and Improving Teaching

III. University Teacher Graduate Interviews
   a. Instructional Strategies
   b. Analysis of Teaching
   c. As a Tool of Reinforcement
   d. As a Tool of Reflective Teaching
   e. Exposure to Integration
   f. Cost Factor or Availability of Materials
   g. Difficulties

IV. University Student Interviews
   a. Mode of Presentation
   b. Classroom Interactions
   c. Effective Strategies
   d. Transfer to Practice

---

teacher education with biology, chemistry, and physics. In essence, the project moved the school of education into closer contact with the arts and sciences, and consequently collaboration between science and teacher education faculty with these other units was noticeably increased.

Major Thrust of Interactive Video. From a faculty perspective, this finding dealt with the fact that the interactive video helped to put content and teaching methods together using a multimedia bridge. Interactive media was not just an opportunity for faculty from different disciplines and colleges to interact more but for students to see the connection of content with teaching/instruction as reinforced through the use of media.

Vanderbilt University Administration Support. This support was a key facilitating factor in the project. It was strong and the Administration including the Chancellor was acknowledged by all faculty members interviewed. That support consisted of start-up grants, cost sharing, encouragement of faculty outside teacher education to get involved in the project, and participation in dissemination efforts to external agencies.

Measurable Student Outcomes. Vanderbilt University graduates and interns were indeed using inquiry-based lessons in their regular and practicum classrooms with confidence and conviction. They had a positive attitude towards teaching (not directly measured or quantified during the original project).

Transferability to Schools Employing Vanderbilt Graduates. The availability of technological and other instructional resources in schools where Vanderbilt graduates have been employed was not a major determining factor in their use of instructional techniques learned through methods courses associated with the project. (However, interviews with graduates in the field indicated a very different picture of this variable.)

RESULTS (EMERGENT EXPLANATORY THEMES)

The initial data categories were critical building blocks for the next phase of analysis and interpretation. The evaluators began to ask questions about whether there were broader themes that seemed to cut across the data and that would explain what had been observed. A beginning set of themes was generated from the results obtained during the first site visit and were then used to modify the interview questions employed during the second visit. In other words, the beginning themes constituted a form of hypotheses to be explored in a subtle manner on a return visit to Vanderbilt. Would they hold up to scrutiny and would they explain what was being observed? Those themes which came from all sources of data are summarized below. (Also see Table 2.)

Administrative Support was evident throughout the interviews. The project enjoyed a considerable amount of collegial, professional and financial support from Vanderbilt administrators. In particular, as noted before, institutional grants for faculties outside science education to get involved in the project were made readily available. Faculty members from the arts and sciences have been encouraged to participate and develop science courses to provide more hands-on experiences to students who are prospective teachers. In addition, support included release time for faculty to work with the project, funding for dissemination, and fiscal help to hire full-time permanent staff to program and edit project videos and hypermedia products. This support was evident from the University Chancellor and the Provost to the Teacher Education Department Chair.

Osmosis/Permeation. As the name implies, the project's approach to teaching undergraduate chemistry using interactive media has permeated into science education, mathematics education, reading education and social studies education. The philosophy of the project influenced several faculty members of the biology and physics departments to redesign undergraduate courses to incorporate interactive media. Both administrative support and faculty interest are the major reasons underlying this osmosis effect. The academic atmosphere itself became conducive to the project ideas as faculty members excited about the interactive media talked about it in their casual meetings and lunch breaks, began to think of ways to integrate interactive media-based approaches into teaching, and encouraged others to become involved.

Student Perception on Campus. Students felt encouraged to learn methods via the interactive media approach. They saw the connections between science content and methodology with respect to teaching and learning. For example, the project helped students realize why hands-on science should be emphasized at the elementary school level and why understanding of the details of discovery/inquiry learning strategies can help them to become reflective teachers. They enjoyed the group discussions and analysis of contrasting teaching episodes as part of the media based methods courses. The interactive videodiscs had a high degree of flexibility which they appreciated. (Students felt that the media was easy to use with little or no technical assistance.) Through the technology they were able to see and experience the reaction of children to various teaching techniques.

Student Perception in Schools. Vanderbilt graduates supported the idea that the interactive media course influenced their knowledge and understanding of effective ways of teaching science. The video based science methods course enabled them to plan and teach hands-on science with confidence, in addition to relating science with other disciplines and societal issues in classrooms. Conversely, they found it demanding to implement what they learned at Vanderbilt within their classroom settings due to various management issues which were not emphasized in the interactive videos. (Even when equipment was available which was true in many instances, the nature of that availability was in question as was the quality of equipment and its condition. Project graduates may have confused the hands-on discovery methods learned via the interactive video project with the technology itself.)

Technical Support. It was obvious that excellent technical support was continuously present and was a necessary part of the project and the subsequent use of its products. It seems that a full-time programmer and a video editor on staff were, collectively, a key reason for the successful implementation of the project. The technical staff helped the Department of Teaching and Learning in making the project based interactive materials easily accessible in the form of a Compact Disc and a HyperCard stack for student use. It was also clear that technical support was there not only for the researchers, but for interested faculty within and outside of science education. Technical support in the form of quick troubleshooting was noted by students as a reason that they did not get bogged down whenever they encountered difficulties in using the interactive media. The technical staff contributed significantly toward preparing high quality materials used by project faculty in presentations within and outside of Vanderbilt University.

Cross Fertilization. All interviewees, in one form or another, expressed the thought that the interactive media project has enabled the bridging of both content and methods in science. The project has brought the Peabody College of Education and the College of Arts and Science faculty closer together in collaborative efforts such as writing proposals for funding, exchanging expertise, and seeing the importance of methodology in content and in teacher preparation. The university administration and excellent technical staff accelerated the exchange of ideas and interactive efforts.

Organizational Climate. A climate conducive to research, development, and implementation of innovative teaching ideas at Vanderbilt University was a constant factor in the interviews. Cohesiveness and mutual understanding between the faculty and administrators frequently surfaced during the interviews.

Critical Mass. This refers to having a number of excited people working as a team and who, individually or as a team, often demonstrated that excitement to others. If it were just one person working on the project it might not have been successful in a typical university environment. Six faculty members were key players on the project besides the involvement of other project staff and graduate assistants. They were enthusiastic in a very positive way—it was an infectious enthusiasm. They found what they were doing to be useful for their teaching, they were actually using it in their classrooms, and talking about it in faculty meetings and seminars. As a result, they created a critical mass that promulgated the project throughout the department in which they were located.

FINDINGS FROM OTHER TECHNOLOGY PROJECTS IN SCIENCE TEACHER EDUCATION

A review of the literature revealed not a great deal of information about the evaluation of technology implementation in science education, particularly science teacher education. In an ERIC database search, the descriptors "Teacher Education", "Program Evaluation", "Science Education", and "Technology" were crossed and articles/entries for a 20 year period from the present time back were sought. That search yielded only nine possibly relevant sources (3 in 1997, 2 in 1992, and the rest from 1989 and earlier).

At first glance, the lack of an extensive base may seem surprising. Interestingly and pertinent to this discussion is some work we undertook approximately 10 years ago. Then, a number of analyses of the literature related to evaluation in science education produced only a small set of references regarding the evaluation of teacher training programs in science education. Given the consistency of these two literature searches, a point made in the introduction to this chapter seems to be corroborated, i.e., more evaluative studies of teacher training in science education are needed. They would be especially timely given the opportunities that new technologies offer the field.

From another perspective, however, the lack of literature may come from the way labels and key terms are used in articles, papers, presentations, and other entries in the database. Many studies of technology utilization in courses and in pre-service teacher training are conducted and are important from an evaluation standpoint, but are not primarily intended to be evaluations. Some of them are in science education whereas others are carried out in a variety of fields.

For example, a study by Kenny, Andrews, Vignola, Schilz, and Covert (1999) while not focused on the training of science teachers contains many features analogous to the context evaluation of the Vanderbilt project. Kenny and his associates utilized multiple measures some of which were qualitative to examine and understand the nature of what was taking place in their situation. One of their findings (akin to one in the context based evaluation described earlier) is that students may mistake or focus too heavily on the features of the technology rather than the principles being conveyed through it. In other words, there is the potential of unwanted side effect as a result of technology as the medium of curriculum delivery.

Another example comes from the field of science education. Kelly and Crawford (1996) investigated discourse as it occurred in small groups to look at "the role the computer plays in the group context and the ways that this context is shaped by the computer" (p. 693). Even though their study was based on students in high school and its goal was research not evaluation, implications for the evaluation of projects like Vanderbilt's are evident. It is not much of a stretch to think of different variables that could have been studied in both types of evaluation at Vanderbilt. Discourse analysis could easily have been the emphasis of either evaluation. While a full review of the broader literature is beyond the scope of this chapter, the two citations demonstrate that there are sources that have either a tangential or more direct relationship to the evaluation of science teacher training, especially when it relies heavily on the use of technology.

The guidance obtained from some sources, more closely related to science education, are discussed next. Abell, Cennamo, Anderson, and Bryan (1996) reported on a project involving video-based cases in pre-service science education at Purdue University. Their findings focused on such outcomes as Preparing to teach science, Thinking about teaching and learning, and Understanding elementary classrooms. According to the authors, interactive media "has the potential to create a virtual world in which novice teachers can experience and reflect upon problems and practice," and provide "opportunities for pre-service teachers to construct and refine their personal theories of science teaching in a meaningful context" (p. 137). There is a relationship between these findings and several of the variables from the non-traditional evaluation of the Vanderbilt project listed in Table 2 (e.g., Emergent explanatory themes 3 and 4, and Initial data categories Ia, Ib, Ig, IId, IIIa, IIIb, IIId, IVc, IVd).

In another technology based pre-service science teacher education effort, Weinburgh, Smith, and Smith (1997) noted that their early childhood teacher graduates "often find themselves in a climate that does not encourage the use of technology even when it is available" (p. 45). This finding correlates with items Im, In, IIb, IIIg, and IVd in Table 2. And even further, Yager, Dunkhase, Tillotson, and Glass (1995) in a science-technology reform project that used interactive television for in-service teachers observed that "teacher participants became more enthused with reform efforts and how the project could be used to implement the reforms" (p. 21). Conversely, they pointed out the need for more "training and assistance in techniques for utilizing interactive television for delivering science instruction" (p. 22).

After reviewing four interactive media systems in teacher education utilized for instructional design, context for learning, and video quality purposes, Ladewski (1996) noted that "none of the groups seemed to have developed detailed evaluation criteria as part of the initial design and development effort" (p. 194). She concluded that "much work remains to be done in the area of evaluation of interactive multimedia systems—identifying gaps, in setting up researchable evaluation criteria, in gathering user data" (p. 195). According to Weinburgh, *et al.* (1997), and Krueger, Hansen, and Smaldino (2000) teachers must learn how to integrate interactive media into instruction to promote meaningful learning.

CONCLUSIONS

There are a number of conclusions that emerge from the prior, brief discussion of the evaluation of technology-based training programs for pre-service science teachers. The conclusions are primarily but not exclusively drawn from the evaluation of the Vanderbilt project.

One obvious conclusion is that context is important and should receive more empahsis in these types of evaluations. Context provides a unique way to think about evaluation that is frequently overlooked when outcome or process evaluations are conducted. Looking at the products of large scale projects alone are not sufficient to evaluate their effectiveness. Evaluation must also focus on the context in which the project was developed and implemented.

Secondly, it could be argued that to understand what works and what does not in technology use in science teacher education, will require in-depth multiple-approach, and multiple-method evaluations. In the Vanderbilt case, each of the two separate and distinct evaluations were major undertakings in themselves. Each one produced an extensive set of findings and provided useful information for project decisions and for improvement/change. The fact that the findings were significantly different suggests that perhaps multiple ways to attack the evaluation, even though they will be costly to conduct and demanding of the skills and experiences of the evaluators, are necessary and long overdue. Certainly that is a learning that came out of the evaluation at Vanderbilt and one that should be underscored.

Refer to the combined results of the traditional and non-traditional evaluations of the project provided in Tables 1 and 2. They afford a comprehensive picture of the interactive media science teacher education project that could not be gained from either evaluation by itself. In effect, the findings taken together are reflective of a rich, complex, and, to a certain degree, elusive set of variables that are necessary to deliver technology-based training to prospective science teachers. The context type of model approach, which is used to a lesser extent than standard approaches, led to looking at a host of variables that are not usually included in very much detail in most evaluative studies. The variables are both difficult to characterize and operationally define, yet they have great import for the design and implementation of evaluations as well as for the generation of meaningful findings about programs. It could be argued that technology-based innovations in science education will just not work if the context is not operating in a positive, supportive manner.

A number of factors have to come together for a project to be successful. An environment conducive for change and development, strong administrative support and readily available technical assistance must be there. Administrative support, in particular, is essential not only in terms of funding but more from a perception (whether real or imagined) that administrators are committed to the project, they are interested to the point that they maintain familiarity with what is going on and the outcomes that are being achieved, and eventually they will become active disseminators and promoters of the project. Channels of communication have to be developed and remarkably, in the Vanderbilt case, they were still evident two years after the completion of the project.

Other more subtle aspects of the environment lead to diffusion of the innovation beyond its original, small core of developers. Again referring to Vanderbilt, faculty members, administrators, graduate students, programmer, and video editor of the interactive, hypermedia project coalesced into what may be called a "critical mass", that by virtue of its deep belief in the quality of the innovation began to affect the overall context. That mass essentially created a press in the environment for change. Others began to adopt the innovation in their classes because they felt they were not keeping up or were missing out on a good opportunity to improve their instruction. The norms of the department that housed most of the developers became, in a sense, an atmosphere for change, one that encouraged people to change but did not coerce them to do so.

A third conclusion relates to the products/materials/processes being evaluated that their nature should not be overlooked. The products at Vanderbilt, while high in initial developmental costs, were easily used, required only small amounts of money to adopt, had clearly defined objectives/outcomes, and integrated well into established teacher education courses and programs. A more complex innovation could have been equally facilitated by having the presence of a critical mass of developers and a press for change but probably would not have been as successful as this one was. That is, the product either by design or by chance, would be readily adaptable with minimal costs and inconvenience to many other similar contexts.

According to Ladewski (1996) there is an "optimum balance" and a "trade-off" between the actual classroom processes which take place in the real world, and the kind of information that could be stored, edited and reproduced using technology with its attendant limitations. The innovative nature of the interactive video project (the content and organization of the videos) was of high quality. The project was well structured and well thought-out for improving teacher education. The environment and the product interacted to produce the meaningful change observed at Vanderbilt and one that could be achieved elsewhere.

A fourth conclusion is that context evaluations should not be done at the expense of, or to the detriment of traditional outcome evaluations. In this regard, it should be noted that the context, non-traditional evaluation that was our focus corroborated the previous traditional one and collectively the two evaluations made a much stronger case for the short and long term (the more valuable one) impact of the project.

A fifth conclusion is that there is a need to define, in much greater depth and specificity what is meant by context and "contextual" variables. The nature of context, i.e., general features as well as the specific variables that comprise it, were not defined to a substantial degree in the non-traditional evaluation. Questions about dimensions of institutional climate such as how research and development efforts are encouraged, how the reward system affects climate, feelings of collegiality, past and present collaborative work, administrative willingness to adjust schedules to facilitate projects like the one studied, and other questions might and should have been asked. Operational ways of measuring what essentially were the shadowy dimensions of the context construct were not clear, and they will affect the success of all projects/programs that place high value on the use of technology in science education.

Attempts to look at context in terms of various interconnected elements are necessary in order to design a comprehensive evaluation scheme. One such element could be the context that exists within the boundaries of the immediate program itself including interactions between recipients of service (students, parents, etc.) and deliverers of that service (teachers, counselors, etc.). The stress would be on the content and structure of programs and what is directly provided to program recipients. The nature of the classroom context in which interactive technologies are implemented would be of major concern. (Ladewski, 1996). Such a context may incorporate "a number of different aspects of a multimedia system—the multimedia environment itself, how the user is assisted in gaining access to the multimedia environment, and the larger learning context within which the multimedia tool is used. All of the three interpretations seem relevant in describing the learning context provided by a particular multimedia system" (p. 192).

Context could also be thought of as the environment that exists around the program such as administration, external support staff, and the system surrounding the program. The variables include administrative support, communication to the larger system, impact of the surroundings on the project or program, and the impacts of the program on its environment. The last element of context could be viewed as the "evaluation" or "decision-making" context which refers to the flow of information about the formative and summative evaluation of the program.

There is a need to clarify the variables that constitute these aspects of the overall context. They must be operationally defined with the purpose of determining the role that each plays in the success or failure of a particular program. Retrospectively, applying this kind of thinking to the evaluation of the Vanderbilt project would probably have led to a different categorization of findings and revealed other areas needing review. From our perspective, context is not fully defined in many evaluations. Unless the reader of the evaluation report fully comprehends what is virtually an implicit instead of an explicit definition, much confusion and misunderstanding is possible.

A sixth conclusion is that if multiple approaches to evaluation are proposed, eventually it will be necessary to develop ways to integrate them conceptually as well as methodologically for the evaluation of technology use in science teacher education. This may be easier said, than done. Different approaches usually require different methods and often it becomes difficult to have the skills needed to employ multiple methods and, in turn, even more difficult to integrate the findings produced by them. (See Altschuld and Witkin, 1999 for a full discussion of the complexity of the use of multiple methods as well as examples of where multiple methods strategy have been employed.)

The seventh conclusion is that there are no easy or simple answers, no quick fixes as to how to evaluate technology based pre-service teacher training as determined from the other literature sources reviewed in this chapter. Consider just the idea of what variables should be included in the evaluation.

Contextual variables and their complexity present us with quite a challenge and the evaluation by the Vanderbilt staff dealt extensively with knowledge, attitude, and performance variables. So what else could an evaluator do. For starters, a slew of other variables could be examined such as the nature and quality of learner discourse, the integrity of the content included in technology and the degree to which it represents or is isomorphic with reality and actual situations students will encounter, subtle dimensions of student use of the technology especially when many options are available to them, and so forth. Furthermore, the evaluations could also take place at different times in accord with where the program or product is in terms of its life cycle (Figure 1). The evaluations could be done for different purposes, even including reinvention.

The final (and best) conclusion is one of excitement that evaluation of technology-based efforts and research about such evaluations are only limited by our imaginations. It seems that there are almost no barriers to science education evaluators plunging into this type of work. Look back at the seven conclusions just drawn and ask the question how much can we learn about evaluation as the new millenium moves forward. Or perhaps better questions are what must we learn to make the best possible instructional and growth producing world for children and how can evaluation help shape that world? That is one's charge as an evaluator—one must become an active player in guiding the infusion of technology in science pre-service programs. If evaluators don't identify what works and what doesn't and what induces the best supportive environment, then who will?

ACKNOWLEDGEMENTS

The authors would like to acknowledge Ms. Kaye-Ann Grant for editorial assistance. A portion of this chapter incorporates materials published in the *Journal of Science Education and Technology*, Volume 8, Number 1, pp. 55–65, 1999. Copyright 1999 by Kluwer Academic/Plenum Publishers. Reprinted with permission.

REFERENCES

Abell, S. K., Cennamo, K. S., Anderson, M. A., and Bryan, L. A. (1996). Integrated media classroom cases in elementary science teacher education. *Journal of Computers in Mathematics and Science Teaching*, 15(1/2):137–151.

Altschuld, J. W., Kumar, D. D., Smith, D. W., and Goodway, J. D. (1999). School-based educational innovations: Case illustrations of context-sensitive evaluations. *Family and Community Health*, 22(1):66–79.

Altschuld, J. W., and Kumar, D. D. (1995). Program evaluation in science education: The model perspective. *New Directions for Program Evaluation, No. 65*, 5–17.

Altschuld, J. W., and Witkin, B. (1999). *From needs assessment to action: Transforming needs into solution strategies*. Thousand Oaks, CA: Sage Publications.

Barron, L. C., Joesten, M. D., Goldman, E. S., Hofwolt, C. A., Bibring, J. B., Holladay, W. G., and Sherwood, R. D. (1993). *Improving science education: A collaborative approach to the preparation of elementary school teachers*. A final report to the National Science Foundation under grant number TPE-8950310. Nashville, TN: Vanderbilt University.

Berger, C. F., Lu, C. R., Belzer, S. J., and Voss, B. E. (1994). Research on the uses of technology in science education. In Gabel, D. L. (Ed.), *Handbook of research on science teaching and learning*. New York: Macmillan Publishing Company.

Exline, J. D. (1985). *Virginia's long range plan for science education*. National Science Teachers' Association Supplement of Science Education Suppliers.

Exline, J. D., and Tonelson, S. W. (1987). *Virginia's science education program assessment model resource guide*. National Science Teachers' Association Supplement of Science Education Suppliers.

Field, S. L., and Hill, D. S. (1988). Contextual appraisal: A framework for meaningful evaluation of special education programs. *Remedial and Special Education*, 9(4):22–30.

Goldman, E. S., and Barron, L. C. (1990). Using hypermedia to improve the preparation of elementary teachers. *Journal of Teacher Education*, 41(3):21–31.

Grandgenett, N., Ziebarth, R., Koneck, J., Farnham, M. L., McQuillan, J., and Larson, B. (1992). An investigation of the anticipated use of multimedia by pre-service teachers. *Journal of Educational Multimedia and Hypermedia*, 1:91–102.

Kelly, G. J., and Crawford, T. (1996). Students' interaction with computer representations: Analysis of discourse in laboratory groups. *Journal of Research in Science Teaching*, 33(7):693–707.

Kenny, R. F., Andrews, B. W., Vignola, M. J., Schilz, M. A., and Covert, J. (1999). Toward guidelines for the design of interactive multimedia instruction: Fostering the reflective decision-making of preservice teachers. *Journal of Technology and Teacher Education*, 7(1):13–32.

Krueger, K., Hansen, L., and Smaldino, S. (2000). Preservice teacher technology competencies. *TechTrends*, 44(3):47–50.

Kumar, D. D., and Altschuld, J. W. (1999). Evaluation of interactive media in science education. *Journal of Science Education and Technology*, 8(1):55–65.

Kumar, D. D., Helgeson, S. L., and Fulton, D. C. (1994). A study of interactive video use in science teacher education. *Journal of Instruction Delivery Systems*, 8(4):28–34.

Ladewski, B. (1996). Interactive multimedia learning environments for teacher education: Comparing and contrasting four systems. *Journal of Computers in Mathematics and Science Teaching*, 15(1/2):173–197.

Milken Exchange on Education Technology. (1999). *Will new teachers be prepared to teach in a digital age? A national survey on information technology in teacher education*. Santa Monica, CA: Author.

Pollack, R. A. (1989). Generic videodiscs in education and training. *Instructional Delivery Systems*, 2(5):22.

Small, L. (1988, April). *Science process evaluation model*. A paper presented at the annual meeting of the American Educational Research Association, New Orleans, Louisiana.

Stufflebeam, D. L. (1968). *Evaluation as enlightenment for decision-making*. A paper presented at the working conference on assessment theory sponsored by the Association for Supervision and Curriculum Development. Sarasota, Florida.

Vitale, M. R., and Romance, N. R. (1992). Using videodisc instruction in an elementary science methods course: Remediating science knowledge deficiencies and facilitating science teaching attitudes. *Journal of Research in Science Teaching*, 29(9):915–928.

Weinburgh, M., Smith, L., and Smith K. (1997). Preparing preservice teachers to use technology in teaching math and science. *TechTrends*, 42(5):43–45.

Welch, W. (1974). The process of evaluation. *Journal of Research in Science Teaching*, 11(3):175–184.

Yager, R. E., Dunkhase, J., Tillotson, J., and Glass, R. E. (1995). Science-Technology reform via distance education technology. *TechTrends*, 40(5):19–22.

CHAPTER 8

Evaluation of Science Teaching Performance through Coteaching and Cogenerative Dialoguing

Kenneth Tobin and Wolff-Michael Roth

Educating science teachers has always had its challenges and to the extent that society has changed markedly in the past century so too have those challenges. At the present time teacher educators face a vast array of problems associated with educating science teachers to practice in a multitude of different contexts. One of these is the evaluation of science teachers and teaching. Evaluation normally involves an outsider to the classroom applying a set of culturally and historically developed criteria during a specially arranged event that constitutes the observation of teaching. The event often incorporates additional "anecdotal" information together with artifacts provided by the teacher and/or other stakeholders. For example, teachers may

Kenneth Tobin, Graduate School of Education, University of Pennsylvania, 3700 Walnut Street, Philadelphia, PA 19104-6216.   Wolff-Michael Roth, Applied Cognitive Science, University of Victoria, MacLaurin Building A548, Victoria, BC, V8W 3N4.

Evaluation of Science and Technology Education at the Dawn of a New Millennium, edited by James W. Altschuld and David D. Kumar, Kluwer Academic / Plenum Publishers, New York, 2002.

provide as evidence of their teaching effectiveness videotapes that show teaching in different contexts, reflective journals that include evidence of reflective practices, and portfolios containing artifacts of work selected to show the achievements of their students. Similarly, school administrators might submit evidence to support the teachers' claims, such as written records of formal evaluations undertaken in previous semesters and a summary of the results of a survey of students' perceptions of their learning environment. These artifacts, which can be weighed to inform a final decision about teaching performance, reflect implicit and non-articulated theories and assumptions such as the context-independence of pedagogical knowing. Equipped with a variety of data, an evaluator examines various aspects of classroom life and compares teaching performances to established benchmarks.

When it is conducted in this way, evaluation is an activity that involves judgments from the side using criteria considered salient (by the evaluator) in the setting. Evaluations like these, although well intentioned, might be unfair to teachers and may have little impact on learning to teach. Even though evaluators may be sincere and secure in their knowledge, we regard the security as false. How can it be known that particular teacher actions will produce postulated social and learning outcomes? Or, how can evaluators know if actions they recommend are feasible in the unfolding circumstances of a teacher's praxis? Even in the best of circumstances the recommendations of outsiders can only be regarded as possible narratives for what might have occurred or what might be desirable. In most circumstances, evaluation can be regarded as a form of oppression: senior and powerful others indict and prescribe for actions with an expectation that their exhortations will be followed. Although evaluators' renditions take into account their knowledge and expertise they include only distant and vicarious experiences of teaching and learning, which are the objects of evaluation. When evaluations from the side take place it is not clear to us how an evaluator can determine whether given practices are or are not appropriate in the contexts that apply. Unless evaluators participate in practice, how can they get a sense of the context and even more to the point how can they viably assess whether particular practices will or will not afford the learning of students? How do science teacher educators take account of context when a new science teacher is evaluated[1] and certified to teach? Are there alternatives to models of supervision that involve making evaluations from the side? In this chapter we explore issues

---

[1] We use the term "new teacher" in preference to "student teacher" or "teaching intern," both of which tend to lessen the value of individual as a legitimate participants in school and university activities.

associated with performance assessment in a context of teaching science in urban high schools.

The remainder of the chapter consists of six sections in which we lay out a different practice for teaching and assessing teaching performance. We begin by articulating the theoretical concepts that frame our practice of coteaching/cogenerative dialoguing. A second section presents a description of the site for the research presented in this chapter. In the third section, we present three scenarios from our research database in which new teachers learn to teach science and supervising professors evaluate by coteaching in urban high schools. We then use a metalogue between the two authors—reflecting the way in which we build theory through cogenerative dialoguing—to analyze and interpret the three scenarios from the previous section. Our analysis is conducted in terms of theory and research associated with learning to teach, assessing teacher performance and the context of urban schooling. A fifth section contains a vignette that illustrates some of the roles of an evaluator who participates as a coteacher in an urban science classroom. A metalogue between Michael and Ken identifies salient issues about evaluation, learning to teach and affording the learning of high school students. In our final section, we present implications and conclusions based on this and other research in which we have been involved.

COTEACHING AND THE ASSESSMENT OF TEACHING

Past theories of teaching have not paid sufficient heed to the fact that teaching is something being done rather than being a static set of procedural and declarative knowledge that somehow is transformed and applied to yield classroom actions. We do not see how it is possible for such transformations to occur. What is known and can be written or spoken can serve as referents for reflection and associated planning of intended actions. However, the knowledgeability (know-how in action) that characterizes teaching is ontologically different from what can be spoken and written. Furthermore, there is a precarious relation between plans and situated actions so that what is intended is only partially related to what actually occurs. In praxis, all sorts of things occur and actors can be conscious about some actions and unconscious about others. To eschew the reductionist approaches of previous theories of teaching, we have elaborated a framework that focuses on the phenomenological dimensions of teaching experience, and particularly on those aspects of teaching that resist description (Roth, Lawless, & Masciotra, 2001; Roth, Lawless, & Tobin, 2001; Roth, Masciotra, & Boyd, 1999; Roth & Tobin, 2001b). Central to our framework, which we refer to as praxeology (Gr. *praxis*, action and *logos*, talk) or talk about action, are the notions of habitus and being-in/with.

Habitus and Being-In/With

We propose an epistemology of teaching as praxis as an alternative approach to understanding teaching and the assessment of teaching performance. Our theoretical framework draws on phenomenology and elaborates a praxeology that hinges on *habitus* and *being-in/with* to conceptualize learning to teach and assessing teacher performance through *coteaching*.

Theories of knowing and learning grounded in phenomenology are concerned with everyday praxis and presuppose *being-in* the world as a fundamental condition of all knowing, a non-thematic, unreflective but concerned absorption in everyday activity (Dreyfus, 1991). The world is comprehensible, immediately endowed with meaning, because we have been exposed to its regularities ever since we first entered the world at birth. Active participation with others in our social world opens us to material and cultural conditions that can catalyze changes in the ways we construct others and our own identities (Giddens, 1991). Being-with others in particular social spaces allows us to acquire *habitus* (Bourdieu, 1990, 1997), systems of dispositions for perceiving and interacting. *Habitus* is not directly accessible and describable but reveals itself in practical situations and in the face of practical decision making. *Habitus*, a generative mechanism that produces practical actions, is not fully accessible to our consciousness and therefore remains beyond reflection as it generates the patterned ways we interact with the world (i.e., the practices that embody actions, perceptions, and expectations). Because *habitus* is formed by the regularities of the world, it anticipates these regularities in its conduct and thereby assures a *practical* comprehension of the world that is entirely different from the intentional and conscious decoding acts normally attributed to comprehension.

Habitus is not part of a deterministic system that is static and closed, but is an open system of dispositions that is under continuous experience-dependent transformation. That is, habitus changes through participation in practice. Practical experiences either reinforce or modify existing *habitus* such that it can sustain more viable practices in a particular context. Occasionally *habitus* cannot adapt sufficiently to a social milieu, which reveals itself as practical failure. In such circumstances subsequent practices will likely be more deliberative until the context changes sufficiently for the *habitus* to afford the goals (or until a new and appropriate *habitus* is forged). Importantly, *habitus* can "also can be transformed via socio-analysis, i.e., via an awakening of consciousness and a form of 'self-work' that enables an individual to get a handle on his or her dispositions" (Bourdieu & Wacquant, 1992, p. 133). Thus, reflection is an additional, though not principal mode by which *habitus* is formed and transformed. Our model of coteaching/cogenerative dialoguing takes advantage of both tacit and reflective modes in forging new habitus.

Coteaching and Cogenerative Dialoguing

Informed by activity theory, which is concerned with the irreducible societal nature of human activity, coteaching makes student learning its primary goal. In coteaching, two or more teachers teach at one another's elbow in order to facilitate student learning (Roth & Tobin, 2001a). The coteachers assume collective responsibility by teaching together, at the same time, rather than dividing up tasks to be done independently (divide and conquer). At the same time, coparticipation allows coteachers to learn from each other without making knowing or learning thematic. As the events of the classroom unfold there is more than one teacher to deal with emerging problems in ways that afford the learning of students. Our research conducted in and as part of teaching praxis evidences significant learning of all participants including teachers, evaluators, and researcher-teachers. Interestingly enough, this learning often occurs in unconscious ways and teachers realize only much later what and how much they have learned while working together with one or more colleagues. Even so, it is probable that most of what is learned remains beyond consciousness (Bourdieu, 1997; Dewey, 1933; Giddens, 1984; Lakoff & Johnson, 2000).

Coteaching experiences are coordinated with meetings during which coteachers and students debrief, make sense of the events, evaluate what has happened, critically reflect on their understanding, and construct local theory and new action possibilities. We call this activity *cogenerative dialoguing*, for all participants have equal opportunities to contribute to the construction of evaluation and theory. In an earlier form of coteaching, the teachers involved met later in the day (after class, during recess, after lunch, at the end of the school day) to debrief the shared experiences. In our recent work, (ideally two) students also participate in these meetings to make sense, understand what has happened, and construct generalizations and expand action possibilities (e.g., Tobin, Roth, & Zimmermann, in press). We are interested in open theory, that is, theory constructed by all participants, because theory constructed in this way can be tested by the participants and ultimately leads to change of praxis. In order to guide our meetings, interactions, and types of issues to be addressed, we developed a heuristic (Table

Table 1. Heuristics for Productive Cogenerative
Dialogue Sessions

1. **Respect** (Between participants)
2. **Rapport** (Between participants)
3. **Inclusion of stakeholders** (Student teachers, students, school personnel, high school students, university personnel)
4. **Ways to participate**
   1. Coordinating discussion
   2. Listening attentively
   3. Initiating dialogue/ideas
   4. Posing critical questions
   5. Providing evidence
   6. Expressing an opinion (agree/disagree)
   7. Speaking freely
   8. Clarifying and elaborating on ideas
   9. Suggesting alternatives for actions
   10. Evaluating ideas and practices
5. **Opportunities to participate**
   1. Contributing to an equitable playing field
   2. Listening attentively
   3. Making space to participate
   4. Showing willingness to participate
   5. Making invitations to participate
   6. Refusing all forms of oppression
6. **Discussion topics**
   1. Learning to teach
   2. Teaching and learning
   3. Curriculum
   4. Teaching kids like us
   5. Coteaching
   6. Transformative potential of activities/curriculum
   7. Links to particulars
   8. Quality of the learning environment

1) that is made available to all participants in cogenerative dialoguing so that they can orient their own actions in appropriate ways.

The praxis of coteaching/cogenerative dialoguing necessitates new ways in which traditional roles of new teacher, (cooperating, regular) teacher, supervisor, and evaluator are understood and enacted. Because the approach is symmetrical in the classroom, traditional forms of critique by one individual (e.g., supervisor) of another individual (e.g., new teacher) lose legitimacy. For example, in a traditional situation a supervising university professor might admonish a new teacher for not maintaining high levels of participation. In coteaching, if the supervisor/evaluator observes a lack of participation or notices that a transition wastes valuable learning time, s/he is obliged (an obligation that is shared and therefore socially mediated within the coteaching group) to act. If the action was conscious and of salience to the learning of students then it could be discussed in the cogenerative-dialoguing session.

In our praxis, only insiders contribute to the generation of theory. This approach harbors dangers in that existing understandings within the group of coteachers can be reified and thereby become ideology. That is, when we use only immediate descriptions of the context, we are likely to remain stuck and reproduce ways of perceiving and acting in a particular context. So, while we need our immediate experience in terms of the concepts that correspond to them, we also use cogenerative dialoguing as a means for engaging in critical analysis to come to an understanding that makes salient the fundamental structures of the condition that we are finding ourselves in. This critical analysis requires our personal understanding of praxis and also the expression of a "radical doubt" (Bourdieu, 1992) or "suspicion of ideology" (Markard, 1993) to address the possibility that we remain ideologically stuck in our current understanding.

Performance Evaluation

Praxis has its own constraints; most importantly, praxis unfolds in time and therefore has local coherence whereas evaluation from the outside seeks global coherence, which is nearly always irrelevant for practical action. Practical wisdom is concerned with the appropriateness of actions in the here and now of a situation, not in the coherence and generalizability of actions across different contexts. We therefore propose that all individuals—new teacher, (regular, cooperating) teacher, researcher, supervisor, or evaluator—participate in teaching so that they share symmetrical mediating relations between students and their knowing. There is no time out for practitioners, forcing them to enact their knowledgeability rather than standing back, as theorists and outside evaluators are able to do, to consider and elaborate all (theoretically) possible forms of action (e.g., Bourdieu, 1990; Roth & Tobin, 2001a). These constraints of the temporality of praxis and the praxis-related limitations in a participant's room to maneuver are available only to coparticipants in praxis. Because *habitus* only reveals itself in praxis and in reference to particular situations, what has to be done cannot be pre-specified in the abstract (e.g., in the form of advice to a new teacher for building rapport), but emerges from the contingencies and temporalities of each situation. An assessor can therefore only access *habitus* through coparticipation with the teachers who are being assessed. Hence, coparticipation is not only a fundamental condition for learning to teach, but also for enacting roles as supervisor and evaluator. That is, as a teacher collective, we do not condone the involvement of individuals who construct themselves as outside observers *looking at* teaching and learning rather than *participating in* it.

DEMOGRAPHICS

City High School[2] has about 2,300 students arranged in 10 small learning communities (SLCs) that can be considered as schools within schools. Each SLC has its own students, classrooms, and teachers. The idea is that students stay together with teachers over the duration of their high school lives and greater personalization of the curriculum leads to higher achievement and a feeling of *esprit de corps* as a result of belonging to a SLC and getting to know the students and teachers. Ninety-seven percent of the students at City High are Black and 87% are from low-income families. The percentage of students graduating in four years from City High is 43%. Two of the 10 SLCs have a curriculum intended to be college tracked whereas most of the others are career-oriented. For example, students in the *Health* SLC often pursue health-oriented themes in their studies and they regularly undertake field studies in health institutions. Most of the students in *Health* are female. In contrast, most of the students in *Science, Education and Technology (SET)* are male. The research described here took place in *SET*, although one of the new teachers, Lisa Gray, also taught a class in *Health*.

Coteaching

Coteaching, the approach to teacher education that we describe in this paper, seeks to change the roles and practices of the key stakeholders associated with learning to teach. Briefly stated, during their yearlong internship, most of our teacher-education students coteach with one or more partners, normally including peers and cooperating teachers, and sometimes university supervisors and researchers.

New teachers are encouraged to learn to teach by teaching. They begin to teach almost immediately—not to take over the control of an entire class but to teach at the elbow of the regular classroom teacher. We envisioned a peripheral (yet legitimate) participation in teaching and acknowledged from the outset that coteaching could be arranged differently in different places. In the conversations with prospective teachers and

---

[2] All names for persons and places are pseudonyms (except for the authors).

coops we made it clear that coteaching involved teaching with another and that there were probably many ways to do this. Since the new teachers were to be in schools for an entire year there was the potential for different modes of coteaching to emerge over time.

When university supervisors visited a class we encouraged them to coteach with new teachers and their coop rather than to effect evaluations from viewing teaching from the side or back of the classroom. In this way, they too participated as teacher, to become a teacher in the class with another teacher and in so doing, to build new habitus and an associated room to maneuver (Roth, Lawless, & Masciotra, 2001). The role of the supervisor was not so much to judge from the side as to coteach as part of a collective that assumed responsibility for the learning of the students and to facilitate cogenerative dialogue between coparticipants from with/in.

ASSESSMENTS OF TEACHING

This section contains three scenarios that highlight aspects of teaching and the assessment of teaching performance. In the subsequent section we interpret the scenarios and identify critical issues in a metalogue.

TODAY THEY WERE TERRIBLE

Lisa: I'm so glad you didn't come today. They were terrible. Even Donna was angry with them. For some reason the students didn't want to work.

Ken: Oh. I'm so sorry to hear that. I know exactly how it feels. Sometimes these kids are just so unpredictable. I must say that I have seen Donna struggle with them before, but it is a rare event.

Lisa: We tried to talk to them about it. But they didn't even want to do that. In the end we made it through, but it wasn't pretty.

Ken: Don't be too hard on yourself. The learning comes from the struggle. I am sure you learned a lot.

Lisa: (*shrugging*) Maybe so. But I'm not sure what . . .

Ken: Tomorrow I'll be there for sure and I'll do my best to help and learn from what happens.

Lisa, a new teacher enrolled for a masters degree leading to certification to teach high school biology, was coteaching a biology elective with Greg at City High. The class was small and consisted of 11 grade 9 students (7 male and 4 female) who needed an additional science course to meet the four-credit requirement for graduation. All students were of African American origin. Even though the course was elective the students had not opted to take it. They had been assigned to the class. Concurrently with teaching, Lisa attended a science methods course. Donna was the methods instructor who had extensive experience of teaching science in urban high schools and had several years of experience of teaching at City High with students like those in this class. However, Lisa and Greg, in their second semester of teaching science at City High, had more experience with these particular students and had taught several of them in the previous semester.

The class was unusual in several respects. Donna, as a methods instructor and researcher, was coteaching with Lisa and Greg four days of the week and no cooperating teacher was assigned to supervise them or to assume responsibility for the class. Lisa and Greg were regarded as new teachers who would learn to teach by coteaching during this and two other 90-minute classes each day. Because of the difficulties we all have experienced while teaching at City High, Donna had agreed to collaborate with Lisa and Greg to develop a curriculum that took into account students' interests and life experiences. Lisa and Greg assumed full responsibility for planning and enacting the curriculum; but, Donna collaborated to the extent possible and conducted research on the teaching and learning of science. Lisa and Greg also undertook action research on their own practices, the lifeworlds of their students, and the manner in which the curriculum was enacted. As the university supervisor for Lisa and Greg, Ken visited the classroom on a regular basis and participated in coteaching. In his fourth year of research at City High, Ken was well aware of the difficulties of teaching in ways that were engaging for learners while affording their learning (e.g., Tobin, 2000).

Ken also was aware of the gap between the praxis of teaching and descriptions of praxis as provided in many methods courses. With Michael, he had undertaken research on learning to teach in urban high schools. Together they had explored the manner in which roles of the new teacher, the university supervisor and the methods instructor had adapted to enhance the learning of the high school students while the new teachers learned to teach science (Roth & Tobin, 2001a). One aspect of the research undertaken by Ken and Michael involved the development of a collective responsibility for the learning of students in lessons in which coteaching occurred. Accordingly, the supervisor's role as an assessor of teaching was adapted to be consistent with the idea that all participants, including new teachers, supervisors, methods instructors and high school students would contribute to the collective responsibility for the quality of the learning. We acknowledged that as the curriculum was enacted, expertise was distributed in that any one coteacher was more expert than others were in some situations and some of the time and that his/her role would be correspondingly more central as the lesson unfolded. Yet, through coparticipation in all activities we expected students to learn science and all coteachers to learn from one another about science teaching.

Dealing with Crises

I [Ken] arrived just as the lesson was about to commence. Lisa was looking somewhat flustered and Greg was nowhere to be seen. "He is rounding them up," remarked Lisa with a laugh as she interpreted my quizzical look. Slowly the students arrived in class, some under their own steam and others ushered in by Greg. Donna shuffled papers on the side of the room as Lisa spoke to her about the general approach to be adopted in this lesson. Following the dysfunctional nature of the previous lesson Lisa and Greg had decided to ask students to sign a statement about adhering to school rules and discuss with them the consequences of not conforming to the rules. I was not quite comfortable about what she proposed to do, but I knew from my experience with students at City High that it was best to follow the suggestions of those who have been most closely associated with the students. Accordingly, I resolved to support whatever Lisa and Greg endeavored to accomplish.

Lisa distributed the forms to the students and immediately they were on the offense. With few exceptions the students were opposed to signing the form and they were very vocal in their opposition to the coteachers' request. Two male students were most vocal and aggressive in their opposition. Lisa drew a line in the sand. If the students did not sign they would be taken to the coordinator of the SLC and would not be permitted back in the class until they agreed to abide by the school rules. "If they are school rules why do we sign?" Bobby was convinced that being asked to sign was disrespectful and he was prepared to accept any consequences for not signing. Meanwhile the argument in class was getting more heated. Lakia enjoyed science and was shouting at the males to stop their protests and sign the form so that they could continue their work. In making her point she used obscenities which violated the first rule on the form. Quickly the males pointed out that she had violated the rules and should be suspended. A shouting match quickly developed and things were rapidly getting out of hand. I decided to act. I walked over to Bobby. "Are you going to sign?" "No," replied Bobby defiantly. "Then lets do what we've got to do. Outside!" I motioned to Bobby with a wave of my hand and briskly strode to the door, holding it ajar for Bobby to leave. "Where you goin' man?" he queried. "To see the principal," I stated emphatically so that everyone in the class would hear. "You ain't told me we were goin' to no principal. I ain't goin'" "Oh yes you are! Outside." For a moment I wondered about the sensibility of my confrontational approach. But it was too late. There could be no backing off now. Tracy recognized the drama in the moment and enjoyed this unexpected confrontation. "It's a face off man!" Quickly I moved into the hallway and Bobby followed. "I'm gonna get suspended man!" Bobby was angry with me. "This is your decision Bobby. You are showing disrespect to everyone in there and that cannot happen." "Why you takin' me to the principal man?" "It's your decision," I insisted as Bobby turned to return to the classroom. "If you go back in there you will be in a lot of trouble," I warned him. Bobby turned back to me. "This sucks! Why I 'ave to sign no school rules?" Fortunately I saw a chance to compromise. "If you go back in there you must show respect for your teachers and work without further disruption. OK?" Bobby appeared to welcome a solution in which he did not lose face. He nodded at me and with a scowl returned with me to the classroom.

Ten minutes later I approached Bobby who was silently working. Quietly I asked if he would meet with me at 3 p.m. to talk about the incident. Bobby shook my hand and agreed to meet. I was relieved. Since the lesson was now proceeding in ways that were relatively controlled, I decided to leave to talk with the coordinator of the SLC. As I walked downstairs I saw Greg who remarked "I was just taking Reggie to a non-teaching assistant (NTA). He is always late and we cannot let him in when he is late. We have got to be consistent or they will all come late. He'll be suspended for being late but in the long run this is best for him and the rest of the class. We have got to be consistent." I nodded my agreement and continued my way to the SLC.

Judging from the Back of the Room

Three days after the above visit to the classroom Lisa came to Ken's office visibly upset by the actions of the coordinator of the SLC during a visit to her classroom. It was only the second time in a year that Sonja had observed Lisa and Greg teaching. From Lisa's perspective Sonja lacked understanding of the context for the lesson she observed. This led Sonja to chastise the students and to make unfair judgments about how Lisa and Greg were teaching. Donna, who cotaught with Greg and Lisa, was also disappointed with Sonja's role. The following is an excerpt from an interview with Lisa on the same day that the incident occurred.

> She came in today . . . she went to sit back . . . with Alana to the back of the classroom . . . she did it once before and I normally nod and smile at her when she says things. Because some of the things she says . . . I just don't agree with . . . some of the things she has to say are pretty valid . . . but today it was just like . . . she singled kids out that I thought . . . weren't . . . weren't acting appropri-

ately. She had no . . . she didn't show up . . . she showed up in the middle of the class period . . . after we had an amazing conversation about the difference between brains and computers and the kids were communicating amongst each other and there was lots of cross talk about science. She started as far back in the class as possible so that she couldn't hear what kind of dialogue was going on. . . . she yelled at some students . . . and I was just like . . . so why are you yelling at my students? Why are you undermining my authority . . . that you say I don't really have to begin with? I mean you know the kids that she yelled at already had detention and already were getting a call home and already had a pink slip. So she was basically undermining everything she has already told me that I should be doing. And I mean that's just what kind of ticked me off. I was just like you couldn't . . . you haven't seen the improvement that we've made in the last three days.

She wanted to talk to Greg and I about how we should be dealing with our classroom management. We know. I mean I know I have to start the paperwork process because some of these kids don't want to be there and are undermining what we're trying to do. I know that already. Greg and I are already working on that. And I met with her and I said, 'can we talk now?' And she said, 'I would rather talk with you and Greg both. You know. It is your class.' And I said, 'OK, that's fine.' . . . She said, 'I have some ideas that I really think will work.' I said I probably know already what you're going to say. I'm too nice, I'm not judgmental and I'm giving complements to kids who don't deserve them. Unbelievable! When I think that positive reinforcement is the one thing these kids need. They keep being told that they can't do this and they don't do that and they won't do this . . .

ASSESSING TEACHING PERFORMANCE: LOOKING BACK[3]

Breaking New Ground

Michael: In traditional teacher education, evaluation is a process in which an outside observer makes judgments about a situation that s/he really knows very little about. We might therefore question the claims of such forms of evaluation to be viable descriptions of teacher competence.

Ken: As a science teacher educator I often supervise new teachers. My primary goal in so doing is to ensure that they teach in ways that afford the learning of students. Hence, when I visit the classroom I feel that it is necessary to do more then make a record of what happens and to list the new teacher's strengths and weaknesses. If there is an opportunity for me to act in ways that will afford the learning of the students I do so, based on a

---

[3] According to Bateson (1972), metalogues are conversations in which previous texts and dialogues are brought to a new more general level by abstracting themes from previous accomplishments. Metalogues are reflexive of the dialogic manner in which we make sense as researchers and evaluators.

premise that the new teacher, being in the classroom with me, will learn from the events associated with my practices.

Michael: Your approach really changes the situation, for it makes learning salient as the primary concern of teaching. Furthermore, by participating in teaching you have a better understanding of what can possibly be done in the situation than if you were to look from the outside into this classroom—through a window into the classroom so to speak.

Ken: There are many purposes for assessing teaching performance of new teachers. Being in the classroom with the new teacher allows for coteaching between the supervisor and the new teacher with the goal of improving the learning of the students. Just as a quarterback will often huddle with the offensive team before commencing a play so it is often desirable for the coteachers, including the supervisor, to huddle during a lesson. In this way, coteachers can ascertain that they are all on the same page in their endeavors to enhance the learning of the students.

Michael: What I like about evaluation in the coteaching mode is that an additional person contributes to the learning. In this experience, the evaluator gains a better understanding with the situation, finding out for him/herself what works and what does not. Also, rather than being the subject of your evaluation, which has a tenuous relationship to teacher learning, your coteacher can learn directly from the event. Subsequently, evaluation happens in the cogenerative-dialoguing session but in new ways because all coteachers—cooperating teacher, supervisor, or researcher—and students contribute to the evaluation of teaching and learning. Here, evaluation becomes a collective responsibility for a shared situation. The main question becomes, How can *we* contribute to improve learning and teaching?

Assessment from Afar

Michael: There may be a temptation for Lisa and Greg to deal with the contradiction of having to perform for Sonja by setting aside their concerns for student learning and switch instead to a demonstration of teaching competencies that are likely to appeal to Sonja. In so doing they can be regarded as working around contradictions by staging an event that looks like authentic teaching but which is focused more on teaching performance than the learning of students.

Ken: Sonja is very well meaning in what she does. She believes in her students and will go to the rack for her teachers too. She was concerned about what had been happening in the biology elective and felt that Lisa and Greg needed her clout to get things right. She seemed to have made up her mind about what to do before she ever came into the class.

Michael: What a difference in Lisa's experience between coteaching with you and the evaluation situation with the SLC coordinator. The old paradigm that evaluation is more objective when it involves fly-on-the-wall type observation sits deep. At the same time, this form of evaluation frightens teachers, which I attribute in part to the process of objectification that they experience.

Ken: Sitting at the back of a classroom to formulate discussion points or judgments is not a viable way of knowing about the teaching and learning that is happening in the classroom. The more the assessors distance themselves by sitting in the back or the sides the more it becomes possible to regard the responsibility for the lesson to be with the new teachers. Sitting in the back relinquishes the shared responsibility for affording the learning of students, and to base judgments on decontextualized understandings of what happened.

Michael: Although it is known that the praxis of teaching differs from its representation in theoretical terms, teacher evaluation generally has not been sensitive to the differences between a situation perceived by a distancing observer and by the involved practitioner. Yet practice theorists such as Bourdieu and Holzkamp point out that the situated, temporal affordances and constraints of praxis are available only to the practitioner fully immersed in the situation.[4]

Forging New Habitus

Michael: One of the important issues in teacher evaluation has to be an appropriate conceptualization of the notion of learning. Perhaps too much emphasis is placed on explicit forms of learning and too little on unconscious and unintentional forms of learning.

Ken: Yes. I think that discussions can serve to bring some of the unintentional and unconscious to the fore. Then, as conceptual objects it is possible to use them for creating new understandings grounded in the experiences of teaching and learning and to plan different sets of intentions for the future. Discussions about shared experiences happen at many different times. In all cases I tend to take the perspective that learning to teach always occurs while teaching. Hence no matter what the judgments might be about how dysfunctional a lesson was, there is a potential to learn from every incident that emerges in a classroom. I like to emphasize that learning can be conscious and unconscious, intentional and unintentional. Hence learning by doing is assured while coteaching.

---

[4] Bourdieu, 1997; Holzkamp, 1991.

Michael: Perhaps it would help us understand learning to teach if we focused our attention on the production of knowledgeability, the flexible process of engagement with the world. Knowledgeability, as Jean Lave points out, is routinely in a state of change, involving people who are related in multiple ways, who improvise struggles in situated ways, and for whom the production of failure is also a normal part of routine collective activity.[5] A focus on knowledgeability leads us to a conceptualization of knowing and learning as (changing) engagement in ever-changing human activities.

Creating Collective Responsibility

Michael: A key aspect of coteaching is the way in which collective responsibility shapes our actions as teachers and as learners.

Ken: Discussions about shared experiences can profitably focus on possible solutions to perceived problems. However, an imperative is to enforce the idea that the responsibility for learning is shared and may extend to system factors. Teachers, particularly in urban schools, can redress many of the problems by planning effectively and enacting the curriculum in ways that engage students.

Michael: This means that evaluators, too, no longer observe teaching and learning from the sidelines but contribute to the collective responsibility of providing the best learning environment for the students in this class and for the other coteachers present. From their position as coteachers, evaluators are no more in the position to construct authoritative descriptions of teaching performance than any other coteacher. The very idea underlying the practice of cogenerative dialoguing was to create open theory, that is, theory to which all participants contribute. Only when we theorize our situation together do we have any chance in bringing about evaluation that has the potential for driving changes in the situation that we theorize.

Ken: On the other hand, system-level factors also need to be changed in many cases and it is important that teachers understand that they should not accept full responsibility for what happens in the classroom.

Michael: I agree. Unless we recognize that what we observe at City High are only subsets of actions from a larger set of societally available actions, there is little emancipatory power in our work. If we simply contribute to make new teachers and students fit into an unjust society then we reproduce inequality and contribute to the subjugation of all to middle-class values. In this sense, evaluation can actually be one of the

---

[5] Lave, 1993.

contributing elements to the status quo through cultural reproduction of inequity.[6]

Avoiding Ideology

Ken: Cogenerative dialogues, involving representatives from those who have shared a teaching and learning experience are ideal opportunities to reflect on practice and to infuse into the discussions relevant theory and research. It also is significant that productive discussions ought to be challenging so that speakers should always provide a rationale and justification for perspectives and all participants should show radical doubt about claims made about the effectiveness of particular practices and recommendations for future practice.

Michael: Indeed, with our focus on theorizing from inside the situation, we have to be acutely sensitive to the possibility of getting stuck in ideology and its insensitivity to blind spots, which contributes to the reproduction of existing situations rather than to the production of change and more equitable situations. As researchers, supervisors, cooperating teachers, new teachers, and students, we are located differently in the activity system that focuses on student learning. These different locations provide different horizons for interpreting classroom events and therefore constitute an opportunity for dealing with particular ideologies.

What Have We Learned?

Just what role does evaluation play in the three scenarios? In the first scenario Lisa and Greg came to the conclusion that the lesson was very bad indeed. By that they meant that the students refused to cooperate and actively resisted the coteachers' efforts to participate seriously in the intended curriculum. Lisa and Greg were committed to their roles in an activity system in which student learning was the object of the activity. When the students resisted, Lisa and Greg endeavored to negotiate with them to re-focus their participation in the intended curriculum. Their commitment to facilitating the learning of the students was also evident in their ongoing efforts to build a curriculum around the interests and knowledge of the students. With the assistance of Donna they planned activities that were of potential interest to the students and consistently sought students' feedback on how to better teach someone like them.

---

[6] Bourdieu & Passeron, 1979.

Although neither of us were present at the lesson that is the focus of "Today they were Terrible" we have experienced many lessons at City High that could aptly be described as dysfunctional. Traditional evaluation, such as the one conducted by Sonja, might lay blame with the teachers, our own evaluation focuses on the entire activity system, its historical development, and its societally mediated character. Thus, teaching, as learning and resistance, need to be factored into why intended learning did not happen in this particular lesson.

Clues as to why the students, with few exceptions, were so resistant to the intended curriculum when it so obviously had been designed with their interests in mind can be seen in the concept of "hallway hangers" (MacLeod, 1995). These hallway hangers rejected the dominant achievement ideology, which assumes that hard work and academic success leads to a better life. They had experiences that indeed showed the fallacy of the achievement ideology and preferred instead to shoot pool, drink beer, and smoke pot rather than to participate in schooling. Not surprisingly they failed to graduate from high school and to proceed to university. Our data from City High suggest that a majority of the students reject the achievement ideology and do not take seriously the necessity to graduate in four years and to proceed to a college education. So few do it and those who do are not necessarily better placed with respect to employment and economic resources than those who drop out. Perhaps the students resist the dominant ideology of a system that blatantly advantages white middle-class males.

Although almost 100 percent of the students at City High are African American the school endeavors to enact a curriculum that is similar in form to what might be found in an affluent middle class suburb in which most students are white. Policies designed for all students characterize the school and an enforcement system that is highly controlled. Each morning students are herded through a single entrance and are required to pass through metal detectors. The process is in many ways dehumanizing and a sign that the students are not trusted. Safety is on everybody's mind and NTAs are present in each of the SLCs to control the students. Not surprisingly there are signs of widespread resistance to the dominant "official" school code and evidence of the enactment of alternative codes throughout the school. As Boykin (1986) has pointed out, these students live their lives as minorities in a dominant white culture where their ideologies must be enacted within a context of hegemonic practices that disadvantage them. Schooling is a potential candidate for hegemonic practice. It is well documented that African American youth fail to achieve at the same level as white youth, especially in science. Yet there is widespread acceptance of the hegemony that what happens at school is normal and a matter of common sense—

even though the evidence suggests that what happens at City High is oppressive to African American youth.

It is in this context that Lisa and Greg teach and learn to teach. The foregoing analysis, which made evident the societal mediation of the events at the school and in the classroom, was needed for change does not come about or last unless societal contradictions are being addressed. We believe that our coteaching/cogenerative dialoguing paradigm, which involves other teachers and students in the evaluation of the entire activity system, is a better model than traditional forms of teaching, which merely contributes to the reproduction of an inequitable society. It is through the taking of collective responsibility by coteachers and students that currently existing vicious cycles of failure to learn can be broken.

Ken's role as a science teacher educator, supervisor, and evaluator are potentially problematic as far as his actions leading to social transformation of the students. Like other new science teachers, Lisa and Greg were assigned to teach in pairs and learn to teach through coteaching. All are assigned to neighborhood high schools like City High and teach in career oriented SLCs. Without exception the new teachers are middle class and only one (out of eight) from Lisa and Greg's cohort was African American. Initially the teaching habitus they bring with them to City High, formed in and by their middle-class culture, is not suited to teaching the students whom they are assigned to teach. Accordingly, the process of learning to teach is slow and involves the adaptation and building of new habitus. In this situation, evaluation needs to be situated rather than abstract, in order to support the transformation and change required to develop expertise in teaching in schools such as City High. Furthermore, evaluation needs to address and contribute to change of the entire activity system rather than constructing attributes for individual new teachers irrespective of the situations in which they do their work. Because Lisa and Greg teach in an urban school where resources are scarce, the curriculum usually enacted by the regular science teachers involves the use of outdated books and a focus on learning facts. In fact, for the past four years the science curriculum at City High where most new teachers are assigned, has been defined by the new teachers rather than the regular school faculty who have remained in the school. Thus, a contextual factor that is significant is assessing teacher performance concerns the accessibility of suitable equipment and materials, maintenance, and disposal of wastes. City High, like many urban high schools has scarce resources and few structures to support curricula like those enacted by new teachers.

Through our ongoing research at City High our own ideology is always under close scrutiny. We doubt the wisdom of enacting a curriculum that is focused on the National Standards, which have been formulated as

a set of decontextualized propositions; these standards embody and contribute to the reproduction of middle-class culture and values (Rodriguez,
1997). To bring about change, (self-) evaluation needs to be paired with
action research to enable necessary changes. In this spirit, Lisa studied
the resistance of students at different institutional nodes throughout the
school. Using the concept of culture as a site for resistance and struggle
(Sewell, 1999), Lisa came to understand her classroom in terms of struggles
between dominant and dominated ideologies. Greg also undertook
research. He investigated the impact of poverty on the lives of the students.
Since the vast majority of the students live beneath the poverty line it was
informative to find out how limited access to money affects the lives of the
students.

As the SLC coordinator Sonja leaves her stamp on the entire SLC.
She can create local policies for her school within a school and with the
assistance of NTAs and a former student of the school she has made major
strides in the past several years in terms of creating a stronger academic
focus and a sense of collegiality among the faculty. She is a very strong voice
and does not back off in the face of occasional student violence or in her
conversations with faculty. She has a perspective on what the SLC will be
like and she works with others in a very collegial manner to attain her
vision. When Sonja entered the classroom to observe Lisa and Greg it
seemed as if she already knew what she wanted to advise them. Her reason
for visiting the class was to legitimate her advising Lisa and Greg how to
be better teachers with these students. Her motivations were solid. Sonja
wanted the best for the students and also for the new teachers. However,
her view of teaching was that there is a correct way to teach and she knew
what that is. Whether she was speaking to her veteran teachers or the new
teachers Sonja positions herself as expert, speaks with authority and does
not expect to be challenged. Although Sonja knew about our commitment
to coteaching and saw the faculty, new teachers and supervisors participating in coteaching, she had made no serious effort to coparticipate in this
way. Accordingly, her perspectives were distant and discursive. Her suggestions to Lisa are to gain control in the conventional ways. "Send them out,
put them on work detail, or suspend them." "Don't praise them when their
work level is too low." "Don't be so friendly." "Don't smile so much." These
suggestions are not grounded in a shared experience and stand in stark contrast to Ken, Lisa, Greg, and Donna who readily accepted responsibility for
the learning environments in the classes they cotaught. Sonja's advice is
predicated on Lisa and Greg having the responsibility for the class and the
advice was tailored for Lisa and Greg rather than for the collective "we"
that worked to better the school experience of *these* students who attended
the grade-9 science class. If Lisa and Greg were to receive too much advice

of this type they might begin to play a game to resolve the contradiction of wanting to enhance the learning of science, on the one hand, and teach as Sonja suggested, on the other hand. Teaching as Sonja suggested, in our view, only contributes to maintaining the oppressive status quo rather then affording the learning of science.

EVALUATING TEACHING AND LEARNING SCIENCE

This section contains a vignette that describes Ken's role as a supervisor as he coteaches with Lisa and Greg in a biology class, a field that is outside of his science specialization (physics). The vignette is written in Ken's voice.

Participant Perspective

"Hypertonic or hypotonic?" Lisa approached me with a perplexed look of urgency on her face. I shrugged. "I have no idea," I said, a little embarrassed at not knowing the answer to her question. "My background is in physics." My explanation was unnecessary because Lisa had moved on. A group of students needed assistance in measuring the size of a bloated water-filled egg. Each group of students had removed the shell from several eggs by soaking them in vinegar and then placed them in various solutions (i.e., water, dilute salt solution, and syrup). At various stages of my teaching career I had known what these terms meant and I knew for sure that their meaning related to the concentration of solute in and outside of the egg. I also knew that the soft inner coating of an egg was a semi-permeable membrane. As I searched for a biology textbook I stopped by several groups of students and asked them about hypertonic solutions. There was confusion in making sense of one of the results and it was not clear to me whether the confusion was related to the terminology or to the concentration of salt in the beaker being less than the concentration of salt in the egg.

Greg was totally occupied in activities in which he appeared to be measuring the volume of different eggs. Lisa asked him about the syrup. "Is it hypo or hyper?" she asked. "It's ambiguous. It could be either." Greg smiled as he continued his activity with several students. As I eavesdropped on the conversation I thought about the extent to which Greg was virtually invisible in the class. He seemed to blend in with the students and they all appeared to like him.

The students were grouped and engaged in myriad activities that reflected what I perceived as a better than average day. Even so there was ample evidence of students pursuing social agendas. Alana was flirting with

Kareem and both were distracted from the lab activity. Octavia was cleaning out a syringe over at a large metal sink that was half filled with dirty glassware. Or was she playing? What appeared to be a legitimate cleaning activity soon revealed itself to be an opportunity for her to fill the syringe with water and then use it as a water cannon. At present she was squirting water into the sink and its dirty glassware, occasionally spraying the wall. It was just a matter of time before she would be squirting the water at other students. Anticipating problems I moved in her direction. She saw me coming and with a grin aimed the syringe at me. Instinctively I stopped and reminded her that this was no time for me to take a bath. I smiled and moved toward her as she eluded me by moving quickly toward a group of males, spraying them with water as she evaded my supervision. My mind left Octavia. "Let her do what she has to do," I thought as I returned to the task of locating a textbook.

My mind was on the science-related concepts needed to interpret the lab. The terminology would be a problem for these students. Solute and solvent were easily confused terms and yet both were important in this lab. They needed to realize that the concentration of ions in solution on either side of the membrane was salient. If there were a difference in concentration then the solute would move through the membrane from high to low so as to even out the concentration. But it was the solvent that would do the moving not the ions I reasoned. And some solvents would be unable to move through the membrane at all because of the molecular structure of the membrane and the molecules. As I searched I was figuring out the design of the lab and potential difficulties these students might experience. Finally I located a spare book and using the index quickly found a table containing hypotonic and hypertonic and relating the terms of the concentration of a solution on either side of a membrane. Having assured myself of the difference I returned to my role of moving from group to group to check on what was happening.

The class consisted mainly of African American males. They were sophomores and juniors and therefore were much more easily managed than the freshman class which Lisa and Brian cotaught earlier in the day. The 20 males were arranged in groups of 4–5 students, which were located at large workbenches. Their shelled eggs had been soaking overnight and they had the task of removing the eggs from solution and then measuring various aspects of the physical properties of the egg in order to ascertain what had happened. There were also a small number of female African American students in the class. Two had opted out of the lesson because it was "gross" and they would not touch the eggs. Octavia was circulating the room but mainly focused her activities on the sink where she was presently playing with food coloring. She had covered her hand in red dye and now was pleading with Lisa to let her go to get help since she had cut her hand. Her act was somewhat convincing to me, but not to Lisa who playfully suggested Octavia wash her hands and get back to the lab. Alana was circulating too. I was unsure of which students she was assigned to work with. But she seemed to be in no mood to work for sustained periods. Teesha was serious in her efforts to learn but was a focus for male attention. Like Alana she was playful at times and her good fun response to the advances of males appeared to give some of them encouragement to spend time interacting with her.

I turned toward two female students who were seated at desks and appeared not to be engaged at all. What's up? I inquired as I approached. "We ain't doin' it. It's gross. I ain't touchin' it." "Come on it's no different then cooking breakfast." I felt chastened by my sexist allusion that females might know about eggs because of preparing breakfast. "We ain't doin' it." Arnetha was emphatic. "Then let me show you what it is all about," I said encouragingly. Both ignored me and continued with their social chatter. "Heh, don't ignore me just because I'm old." "Sorry. My bad," said Arnetha with a grin. Neither student moved. "Come on get yourselves to the board. Don't you want to see what is going on? This is cool." As I gave them verbal encouragement I herded them toward a couple of stools close to the chalkboard. They moved and joined me in what turned out to be a mini-lecture for the two of them. I began by reviewing the egg and its basic structure. "It's basically calcium carbonate, just like this chalk," I told them referring to the shell of the egg. When we put it in vinegar the acid will react chemically with the shell to produce a gas and leave a residue that will wash off the membrane of the egg. As I spoke I created a labeled diagram of the experiment and included most of the terms that appeared to be needed and/or that had been used in the earlier part of the lesson.

"Yeeuck!" Eggs were beginning to break in various parts of the classroom as Lisa moved to assist students to clean up the mess. Arnetha was more interested now in the mess that was being created and she was quickly distracted from my explanations. "Riley shoulda given us work to do," she said as she moved away from me. "It's his fault. He shoulda given us somethin' to do." "But it's your education," I protested as I moved with her. "It's you who is just wastin' your time. Why don't you ask him for another activity?" "Nah. He's our teacher. He oughta give us somethin' to do." They moved back to their desks and resumed their socializing as I began to work with some of the males in the class. "Come on Dawoud. Don't miss the chance to learn something." Dawoud and I have known each other through my research activities in the school and he has worked with us as a student-researcher. He looked up seriously and nodded his head in affirmation. For the time I was with the group he worked sedulously on the completion of the task and so did his peer group. But as soon as I moved away they relaxed and continued with their good-natured socializing.

At last Teesha was alone. Seizing the opportunity I approached her. She smiled as I asked her whether she was trying to become a thug. "Oh. My bad." She slid back her hood and waited for me to continue. "So, what do you think about the lesson?" "It fun," she remarked. "You have some good teachers," I suggested and she readily agreed. "How might we make this lesson better?" Teesha surveyed the class and quickly began to offer suggestions. "They not workin'. Riley gotta give them less time." My thoughts exactly. The lab had gone for too long and there was no sense of urgency. "He gotta get them workin'," she said with a wave of her hand. Once again I agreed with her. "Will you talk to Greg and Lisa about the lesson?" Teesha nodded her affirmation and our discussion was over.

"I gotta go to the bathroom." Octavia was remonstrating with Lisa who assured her that: "It won't make a difference. It won't wash off. It won't hurt you. It's in the foods you eat." Octavia looked dubiously at the red blotches over her arms and hands. "Nah. I gotta go to the bathroom." "It'll wear off," I assured her. "You sure?" I nodded my affirmation and moved toward the front left of the class.

Greg was getting ready to wrap up the lesson. He positioned the overhead projector as Lisa facilitated the tidy up of lab materials. The transition was an opportunity for socializing of various forms as some of the males practiced their latest dance moves, good-naturedly sparring with one another and hanging out for a bit. Some of the females also socialized and one or two interacted flirtatiously with some of the males. In due course they were all seated and Lisa and I went from student to student encouraging them to get their notebooks out and open. Greg's expectations were different than ours. He wanted them to watch his explanation of the lab whereas Lisa and I wanted them to have notebooks and pen out and to get down the salient points into their notes. Accordingly, as we moved around the classroom those students with whom we had most social capital complied with our requests and others ignored us, complaining that Riley didn't require them to take notes. "So be it!" My mind switched to Greg's diagrams and explanations.

Greg was focusing his explanations on one anomalous result. He had expected the water to diffuse from inside the egg in the saline solution and was making the point that science does not always turn out as you expect. He made several excellent points about the nature of science and also the experimental method in which the design can be adapted to examine what is going on in a closer way, perhaps in this case by making a more concentrated salt solution. As he explained about hypotonic situations I intervened. "There are so many complicated words here," I said. "Do they know that it is the solvent that moves through the membrane but the concentration of the solute that determines which way it flows?" Immediately Greg began to adapt the diagram on the overhead by adding black dots to show concentration. "Some of them don't know who you are," said Lisa. "Who doesn't know who this gentleman is? Who knows who he is?" Some of the students shouted my name and others raised their hand. "He's my teacher," said Lisa. Some of the students who knew me well explained to others about my role and Greg continued with his explanation, now with a more detailed set of diagrams. I moved toward to door. "I have a meeting," I explained to Lisa as I moved off. "Can you talk to Teesha about the lesson? She has some interesting perceptions that might help us to do a better job." Lisa nodded affirmatively and I gave Greg a nod of encouragement as I left the classroom for the university.

Metalogue

Michael: Your account of this chemistry lesson articulates some of the central features of our coteaching/cogenerative-dialoguing paradigm. Although you are Lisa's methods teacher, you are not merely an observer noting things that work and those that don't. Rather, you participated in the lesson, that is, in fostering learning, and at the same time in coming to understand what works and what does not work. Rather than providing a posteriori advice, the viability of which can never be tested, you actively encouraged reluctant students to participate. In dealing with their resistance, you experienced what was possible or impossible in the situation. But because Greg and Lisa were in the lead, you were able to step back and think about what they were doing. In this way, you were in a position to take a step back and reflect on the ongoing events.

Ken: I evaluate this as another "both/and" day. The lesson was a good one and the coteaching between Lisa, Greg and I was well articulated and to the benefit of the students. Not only that, the lesson provided a nice context for learning to teach. On the other hand, the lesson was as "event full" as most classes in this school. Many students seemed to have goals and participate in ways that I considered to be unrelated to science. But between the three of us we had kept many of them involved in science and there had been no serious problems. I was convinced that Greg needed to plan more activities and break his 90-minute classes into 15-minute chunks, pacing students through each chunk. The "urban shuffle" was as evident today as it ever has been. The students get their work done as they participate in socially focused activities for most of the time.

Michael: Your mention of the "urban shuffle" also highlights the issue that articulating something as a problem and understanding it does not necessarily assist in changing classroom practice. We have talked about this "problem" for quite some time. Yet despite this awareness, and despite your concrete actions designed to deal with the phenomenon, it is still a salient feature of teaching and learning at City High. Your attempt to redress the situation is a neat example of actions coteaching evaluators might take rather than waiting to chastise Lisa and Greg for not having dealt with this phenomenon. Here, you were personally experiencing the difficulties to make changes—difficulties that traditional evaluation would not have made thematic. Advising Teesha, Lisa, and Greg to talk about how to address the urban shuffle ("He gotta get them workin'").

Ken: In saying this I am not being critical of Greg for the lesson and the way it played out. I have taught in this school for three years and know how difficult it is to get the students to work in a sustained way and to learn science of the type that was the focus of today's lesson. There are aspects of what happened that I would change, but these aspects pertained to my own role, not Greg's or Lisa's. My hope is that Teesha will speak to Greg and Lisa about shortening activities and keeping the students focused for shorter periods of time.

Michael: Such involvement of students in the evaluation of teaching and learning constitutes the very core of our coteaching/cogenerative-dialoguing model. It is only through such involvement of all stakeholders that we can hope to bring about changes that allow the African American students at schools like City High to succeed in ways that are consistent with their own values.

Ken: My sense is that Greg might find her suggestions more persuasive than mine. For several months now I have been advocating that all teachers in the school plan at least six activities for each 90-minute period. Neither the teachers nor the students appear to be using the full 90 minutes of activity productively. The students are often cooperative but rarely do I see most students having a sense of urgency about time and its use.

Michael: But this may be more your problem than it is that of the students. I think that we have to look at this "lack of urgency about time" in terms of their own goals, which are not those of middle-class students.

Ken: You are right, perhaps this is my problem. I am aware that African Americans and Caucasians have different perceptions of time and its use.[7] I do have that Western efficiency goal and like to see more purposeful activity in class.

Michael: Yes, but is this goal appropriate and viable to assist African American students from situations of poverty to be successful in their terms?

---

[7] See Boykin, 1986.

Ken: During the "shuffle" they are dealing with their social agendas. I do value the students learning of science and argue that it has transformative potential for their lives. For most students science is just another course they have to pass and the value of graduating from high school might not even be seen as potentially transformative. A desire to pass may be more related to pleasing parents/guardians than to a belief that high-school diplomas provide increased access to wealth or employment. In fact, significant numbers of students have rejected the achievement hegemony[8] and see school as having purposes other than learning science.

Michael: I agree with you on this point. And it is therefore important to involve the students at City High in evaluation of learning and teaching, and to involve them in changing their situation all the while allowing the students to empower themselves. But let me address another issue. An important component of evaluation in the coteaching/cogenerative-dialoguing paradigm is that evaluation is not separate from action. Telling new teachers what they might have done often does not help because the particular situation may not crop up again. Furthermore, advice from the sidelines may or may not be viable. I believe that it is better to see if some action brings about a desired change, and subsequently to talk about the success or failure of the intervention. In this way, the evaluator is an integral part of the situation, and involved in bringing about change in a concrete way. In the relation between evaluation and teaching, I see parallels with the relationship between theory and praxis. I see little utility in uncoupling the two and favor the testing of actions and change in the very moment that we engage in praxis.

Ken: I agree. In one of my first studies on the teaching and learning of science in urban high schools I learned that I should never ask students to do something unless I could deal with the consequences of them not doing as I instructed. Previously, I might have told new teachers to do this or that. By addressing the situation directly, I can actually find out whether what I might have proposed as part of an evaluation actually works then and there when the action is most relevant. Also by my own actions to manage student behavior and participation I have learned not to make matters worse. To the extent possible I endeavor to focus my efforts in the class on facilitating the learning of students who came to learn. In ways that are respectful of the students, I try to get them engaged when they are not participating on task. As I interact with students to assist their learning I endeavor to earn their respect and build rapport.

---

[8] See MacLeod, 1995.

Let me address another issue. When the question about the difference between hypertonic and hypotonic solutions arose, my intention was, since I cotaught, to learn the terminology *then and there* and to assist students in learning what they were expected to know and use.

Michael: Here, I am a bit worried about the perceived need to make students know the difference between "hypertonic" and "hypotonic" solutions. These are just words, which are likely to have no relevance to the lives of the students. Furthermore, even Lisa and you do not know how to use the two words, which does not take away from other aspects of your life. Why do you and Lisa insist on telling students about the concepts when even Lisa and you, highly trained individuals, forget how to use them?

Ken: You make a good point. Teachers need to be selective about what terminology to include in the science toolkit. Knowing some basic terminology of science is important for these students so that they can participate in science-like discussions and build understandings about what they are learning. Not only that, knowing science terms can connect to the students' identities.

Michael: Well, I believe that we have to rethink this. If we want to make science relevant to everyday out-of-school situations, then we have to accept that the monoglossic form of scientific discourse will change. Absorbed into the already heteroglossic everyday discourse, science changes and, in fact, becomes more democratic.[9]

CONCLUSION

In the course of our work at City High, we adopted activity theory (Engeström, 1987; Leont'ev, 1978) and poststructuralism (Bourdieu, 1997) as ways to think of learning to teach and assessing teaching performance. This has allowed us to create an awareness of the advantages of creating a collective responsibility for the students and their learning of science. Simultaneously, we are attuned to the real consequences of competing ideologies and tendencies for the cultural production of students to recreate their locations in social space. Of course being aware is no guarantee that we will succeed in any of our endeavors. However, as we create new roles for assessing teaching performance, by coparticipating in classrooms and by making evaluation a collective responsibility of coteachers and students alike, we come ever closer to the tenets of Guba and Lincoln's (1989) fourth generation evaluation. As supervisors we learn from our evaluation

---

[9] Lee & Roth, 2001.

in that our constructions vary over time, always seeking to be viable in changing contexts of learning and teaching. Collectively, we make every effort to ascertain what is happening and why it is happening from the perspectives of the key stakeholders and then to contribute to educate ourselves with respect to the perspectives of one another. As a result of what we learn through evaluation we seek to catalyze positive changes that are consistent with the goals of the teachers and students of the school; change is enabled because evaluation is a collective effort. Finally, we are tactical in our efforts to help those who have difficulty helping themselves in a school in which oppression is widespread and involves different stakeholder groups.

We have adopted the premise that teacher education is a transformative activity—not merely for the new teachers but equally important for the students and teachers in the schools where new teachers enact their internships. While learning to teach, the new teachers assigned to City High are able to make a positive difference to the lifeworlds of the students and others in the school. Because City High is such a different school to any that new teachers have previously encountered, their experiences are potentially rich learning opportunities. Yet, it is imperative to remember that much of what is learned, the special know-how of teaching, will be unconscious, unintended and not accessible through the use of language. Knowing and learning to teach, as Lave (1993) suggested, are flexible processes of changing engagement with an ever-changing world. If that is the case, and we believe that it is, then experiences need to be structured such that new teachers can participate in the myriad events that unfold at a seemingly alarming rate. What we want to support are learning *situations* and forms of *interactions*, neither of which can be reduced to individualized knowledge and properties of individuals. As teacher educators, we strive to support the emergence of collective responsibility for teaching, learning, and the continuous process of evaluating teaching and learning processes to enhance the participation of others (students and teachers) alike. Coteaching and cogenerative dialoguing seem to be the keys to participating in ways that are measured and enjoyable. It seems unfair to assign responsibility (for teaching or evaluating) to any one individual or set of stakeholders. On the contrary, it seems reasonable for groups of stakeholders to assume responsibility for the learning of students (including the students themselves) and through a collective agency to coparticipate in ways that are mutually beneficial to all stakeholders. Just as learning can be a goal for all stakeholders so too can evaluation and assessment. Cogenerative dialoguing has appeal as a suitable venue for all stakeholders to participate in evaluation designed to afford the goals of the collective.

ACKNOWLEDGMENTS

This research was funded in part by the Spencer Foundation and Grant 410-99-0021 from the Social Sciences and Humanities Research Council of Canada.

REFERENCES

Bateson, G. (1972). *Steps to an ecology of mind*. New York: Ballantine.

Bourdieu, P. (1990). *The logic of practice*. Cambridge, England: Polity Press.

Bourdieu, P. (1992). The practice of reflexive sociology (The Paris workshop). In Bourdieu, P. and Wacquant, L. J. D. *An invitation to reflexive sociology* (pp. 216–260). Chicago: The University of Chicago Press.

Bourdieu, P. (1997). *Méditations pascaliennes* [Pascalian meditations]. Paris: Seuil.

Bourdieu, P., and Passeron, J.-C. (1979). *Reproduction in education, society and culture* (Transl. by Richard Nice). Thousand Oaks, CA: Sage.

Bourdieu, P., and Wacquant, L. J. D. (1992). *An invitation to reflexive sociology*. Chicago, IL: The University of Chicago Press.

Boykin, A. W. (1986). The triple quandary and the schooling of Afro-American Children. In Neisser, U. (Ed.), *The school achievement of minority children: New perspectives* (pp. 57–92). Hillsdale, NJ: Lawrence Erlbaum Associates.

Dewey, J. (1933). *How we think*. Boston: Heath.

Dreyfus, H. L. (1991). *Being-in-the-world: A commentary on Heidegger's 'Being and Time,' division I*. Cambridge, MA: MIT Press.

Engeström, Y. (1987). *Learning by expanding: An activity-theoretical approach to developmental research*. Helsinki: Orienta-Konsultit.

Giddens, A. (1984). Structuration theory, empirical research and social critique. In Giddens, A. *The constitution of society: Outline of the theory of structuration* (pp. 281–305). Cambridge, England: Polity Press.

Giddens, A. (1991). *Modernity and self-identity: Self and society in the late modern age*. Stanford, CA: Stanford University Press.

Guba, E., and Lincoln, Y. (1989). *Fourth generation evaluation*. Beverly Hills, CA: Sage.

Holzkamp, K. (1991). Societal and individual life processes. In Tolman, C. W. and Maiers, W. (Eds.), *Critical psychology: Contributions to an historical science of the subject* (pp. 50–64). Cambridge, England: Cambridge University Press.

Lakoff, G., and Johnson, M. (2000). *Philosophy in the flesh: The embodied mind and its challenge to western thought*. New York: Basic Books.

Lave, J. (1993). The practice of learning. In Chaiklin, S. and Lave, J. (Eds.), *Understanding practice: Perspectives on activity and context* (pp. 3–32). Cambridge, England: Cambridge University Press.

Lee, S., and Roth, W.-M. (2001). Monoglossia, heteroglossia, and the public understanding of science: Enriched meanings made in the community. Manuscript submitted for publication.

Leont'ev, A. N. (1978). *Activity, consciousness and personality*. Englewood Cliffs, NJ: Prentice Hall.

MacLeod, J. (1995). *Ain't no makin' it: Leveled aspirations in a low-income neighborhood*. Boulder, CO: Westview Press.

Markard, M. (1993). Kann es in einer Psychologie vom Standpunkt des Subjekts verall-
gemeinerbare Aussagen geben? [Can there by generalizations in a subject-centered
psychology?] *Forum Kritische Psychologie*, 31:29–51.
Rodriguez, A. J. (1997). The dangerous discourse of invisibility: A critique of the National
Research Council's national science education standards. *Journal of Research in Science
Teaching*, 34:19–37.
Roth, W.-M., Lawless, D., and Masciotra, D. (2001). Spielraum and teaching. *Curriculum
Inquiry*.
Roth, W.-M., Lawless, D., and Tobin, K. (2001). Time to teach: Towards a praxeology of teach-
ing. *Canadian Journal of Education*.
Roth, W.-M., Masciotra, D., and Boyd, N. (1999). Becoming-in-the-classroom: a case study of
teacher development through coteaching. *Teaching and Teacher Education*, 15:771–784.
Roth, W.-M., and Tobin, K. (2001a). *At the elbow of another: Learning to teach through coteach-
ing*. New York: Peter Lang.
Roth, W.-M., and Tobin, K. (2001b). Learning to teach science as praxis. *Teaching and Teacher
Education*, 17(7):741–762.
Sewell, W. H. (1999). The concept(s) of culture. In Bonnell, V. E. and Hunt, L. (Eds.),
Beyond the cultural turn: New directions in the study of society and culture (pp. 35–61).
Berkeley, CA: University of California Press.
Tobin, K. (2000). Becoming an urban science educator. *Research in Science Education*,
30:89–106.
Tobin, K., Roth, W.-M., and Zimmermann, A. (in press). Learning to teach science in urban
schools. *Journal of Research in Science Teaching*.

CHAPTER 9

Evaluating Science Inquiry
A Mixed-Method Approach

Douglas Huffman

"It is not the answer that enlightens, but the question."

Eugene Ionesco

INTRODUCTION

For many science educators inquiry is one of the most critical elements of quality science teaching. The National Science Education Standards place inquiry at the foundation of science education stating that, "Inquiry into authentic questions generated from student experience is the central strategy for teaching science" (NRC, 1996, p. 31). From a historical perspective DeBoer (1991) notes, "If a single word had to be chosen to describe the goal of science education during the 30-year period that began in the late 1950s, it would have to be inquiry." (p. 206) The notion of teaching science through inquiry goes back even further than the 1950s. John Dewey (1910) wrote about inquiry in the early 1900s and argued that science teaching

Douglas Huffman, Department of Curriculum and Instruction, University of Minnesota, 358 Peik Hall, 159 Pillsbury Dr. SE, Minneapolis, MN 55455.

Evaluation of Science and Technology Education at the Dawn of a New Millennium, edited by James W. Altschuld and David D. Kumar, Kluwer Academic / Plenum Publishers, New York, 2002.

placed too much emphasis on the presentation of facts and not enough emphasis on the process of science. There is also a long history of research on impact of inquiry in the science classroom. In a meta-analysis of more than 100 studies on inquiry-based curricula Shymansky,. Kyle, and Alport (1982) found that when compared to traditional curricula, inquiry approaches had significant positive impacts on student achievement, attitudes, process skills, problem solving, and critical thinking skills.

Despite its long history and critical importance, inquiry is a difficult concept to define. Over the years there have been multiple definitions of inquiry and slight nuances in the way it has been used in the classroom. For some inquiry is synonymous with hands-on science, discovery learning, or science as process. The field of science education has used terms like inquiry, investigation, and experimentation without always clearly defining their meaning. Some teachers use a guided-inquiry approach where the teacher leads the students through a series of carefully designed steps. Other teachers use an open or free-inquiry approach where students are encouraged to explore and inquire on their own. The National Science Education Standards state that their vision of inquiry is a step beyond "science as a process" where students learned skills such as observation, inference, and experimentation. (NRC, 1996, p. 105) The new vision includes the processes of science, but goes beyond these skills and attempts to combine processes and scientific knowledge.

The different definitions of inquiry and the different ways inquiry is used in the science classroom make it a difficult concept to evaluate. Inquiry is a complex multi-dimensional concept that requires a complex evaluation approach. The purpose of this chapter is to propose a mixed-method approach for evaluating inquiry and provide examples from my own evaluation experience on how to evaluate inquiry in the science classroom. As the foundation of the science classroom it is essential that educators be able to evaluate the impact of inquiry on students, teachers and classrooms alike. Evaluating science inquiry can help not only provide information to those engaged in science education program evaluation, but also help educators better understand the impact of science inquiry and to provide classroom science teachers with critical information to improve their practice.

PART I: DEFINING INQUIRY

Historical Perspectives of Inquiry

In order to define inquiry, it is helpful to look back and consider historical perspectives on inquiry-based science teaching. The following overview of the history of inquiry includes some of the key events and people who have helped to shape current definitions of inquiry. A more complete description is provided by DeBoer (1991).

In the early 1900s The Committee of Ten headed by Charles W. Eliot, then President of Harvard University, made major contributions to promoting a laboratory-based approach to teaching science. Although the view put forth by this committee was not quite inquiry as we know it today, the laboratory approach to teaching science was in stark contrast to the memory studies, and the presentation and recitation of scientific facts that dominated science instruction at the time. According to the Committee of Ten (1893), science education should be based upon lab experiences. The committee did not advocate for students to discover science on their own, but instead believed that students should be carefully guided through lab activities to reach scientific generalizations.

Two of the chairmen of sub-committees formed by the Committee of Ten elaborated on the role of laboratory-based instruction to teach science. Smith and Hall (1902) advocated for an approach to teaching chemistry and physics that emphasized student discovery, developing meaningful understanding of concepts rather than memorization, and the use of practical applications from everyday life. The critical feature of this approach was that it was designed to help students raise interesting questions and learn how to investigate and find answers on their own. The laboratory was expected to help students not only develop observation and reasoning skills, but in addition to help students better understand chemistry concepts.

John Dewey also played a critical role in conceptualizing the role of inquiry in the science classroom. Dewey argued that science teaching placed too much emphasis on the accumulation of facts and not enough emphasis on the method of inquiry. According to Dewey (1910), "Only by taking a hand in the making of knowledge, by transferring guess and opinion into belief authorized by inquiry, does one ever get a knowledge of the method of knowing" (p. 125). Dewey also defined several goals of teaching science as inquiry including thinking and reasoning skills, habits of mind, science content, and understanding science processes (Dewey, 1938).

In the late 1950s Joseph Schwab (1958; 1960) took the notion of inquiry (or enquiry as he referred to it) even further and defined for the field of science education the teaching of science as inquiry. Schwab especially took issue with the presentation of science as a set of facts or irrefutable truths. Schwab referred to this type of teaching as the "rhetoric of conclusions" and believed that it contradicted the nature of science itself. Science is not a set of stable facts, but on the contrary, is a field where ideas are altered in the face of new evidence and new ideas. Schwab noted that most science teaching was inconsistent with the dynamic nature of science, and suggested that if science was taught as inquiry and if students engaged in inquiry themselves it would more accurately reflect the nature of science itself. Schwab defined three different levels of openness that teachers should consider when using a laboratory activity. At the simplest level, the lab poses questions and suggests methods for the student. At the next level, the lab poses questions, but the student must select the appropriate method. At the most open level, the students generate the questions, devise methods of investigation and propose explanations based on the evidence they collect. Schwab basically advocated for more open-ended laboratory activities where students inquire on their own. Schwab also advocated for "enquiry into enquiry" where students examine the evidence and conclusions reached by scientists. In this approach, students are encouraged to discuss alternative explanations, debate the interpretation of data, and closely examine conclusions. Schwab was extremely influential in providing the theoretical foundation for the curriculum development movement that began in the 1950s and 1960s.

Another influential person who provided a foundation for the curriculum development movement of the 1960s was Jerome Bruner. Bruner (1960) advanced the notion that science teaching should emphasize the structure of the discipline rather than just the facts and conclusions. This would result in students learning how different ideas and concepts were related. Bruner also advocated for learning the discipline in the same way that scientists understood the subject. This included not only an in-depth understanding of the big ideas and how they relate, but also an understanding of the methods of inquiry. According to Bruner (1960), the accepted practice of emphasizing facts and conclusions did not allow the students the opportunity to experience the "discovery" of science in the same way a scientist does. In large part the ideas of Jermone Bruner including using discovery learning and teaching science concepts through inquiry was a significant aspect of the science curricula developed in the 1960s.

In the 1960s, the U.S. engaged in one of the most ambitious curriculum reform movements in its history. In response to the Soviet Union's launching of the satellite Sputnik and the fears that the U.S. educational system was falling behind other countries, the National Science Foundation funded numerous curriculum development projects. The result was a major influx of curriculum materials designed to help teachers use an inquiry approach to teach science. For example, curricula developed by the Biological Science Curriculum Study (BSCS), Elementary Science Study (ESS), and Science a Process Approach (SAPA) were all created in this era. Many of the post-Sputnik approaches to inquiry stressed science as a process and specific process skills. Students were encouraged to develop skills such as observing, classifying, measuring, inferring, predicting, forming hypotheses, controlling variables and experimenting. One of the major criticisms of the early inquiry curricula was that they emphasized process skills at the expense of content. While the curricula were definitely engaging for students, some educators were concerned that there was too much emphasis on process skills and not enough on learning science concepts. Ideally, process skills are learned in the context of science content because in many ways process skills cannot be separated from science content anyway. For example, just because a student can make an observation in physics does not necessarily mean that they can also make an observation in biology. These two observations require different science content knowledge. Hence the notion that science teachers should focus on teaching both general skills of observation and observations skills within the context of specific science content.

In the late 1970s the National Science Foundation supported Project Synthesis (Harms & Kohl, 1980). As part of the project, Welch, Klopfer, Aikenhead and Robinson (1981) examined the role of inquiry in science education. They found that the field of science education was using the term inquiry in a variety of ways. To help clarify this state of affairs, Welch *et al.* (1981) attempted to define the desired and the actual state of inquiry in science education. They defined inquiry to be a "... general process by which human beings seek information or understanding. Broadly conceived, inquiry is a way of thought" (p. 33). They divided the domain of inquiry into three main themes: 1) science process skills 2) general inquiry processes and 3) the nature of scientific inquiry. Science process skills included the usual range of skills such as observing, measuring, and interpreting data. General inquiry processes included broader skills such as problem solving, logical reasoning, and critical thinking. The nature of scientific inquiry focused on the epistemological characteristics of inquiry such as the rules of evidence, the structure of knowledge, and assumptions about the natural world. The definitions provided by Project Synthesis helped not only to clarify the term inquiry, but to identify further the extent to which inquiry was being used in the science classroom. As one might expect, Project Synthesis found that despite the emphasis placed upon inquiry by the reform curricula, the actual use of inquiry in the science classroom was minimal (Welch *et al.*, 1981).

In the 1990s the National Research Council worked to develop the National Science Education Standards (NRC, 1996). The Standards were developed over several years and were based upon the input of science teachers, scientists, and science educators. Included in this document was a definition of inquiry. The Standards used the term inquiry to refer to both the ability to engage in inquiry, and to understand about inquiry. According to the Standards:

> Inquiry is a multifaceted activity that involves making observations; posing questions; examining books and other sources of information to see what is already known; planning investigations; reviewing what is already known in light of experimental evidence; using tools to gather, analyze, and interpret data; proposing answers, explanations, and predictions; and communicating the results. Inquiry requires identification of assumptions, use of critical and logical thinking, and consideration of alternative explanations. (p. 23)

The vision of inquiry in the Standards was specifically written to help educators move beyond "science as a process" in which students learned individual process skills such as observing, classifying and predicting. The intent was to encourage science teachers to use inquiry to combine both science process and scientific knowledge. In other words, teachers were encouraged to use inquiry skills to help students develop an in-depth understanding of science. The Standards helped the field of science education move beyond the old process versus content debate and come to terms with the notion that both are critical to developing scientifically literate students. In addition, the view of inquiry presented in the Standards does not refer to one "scientific method" that is characterized by a rigid sequence. Instead, the Standards refer to a logical progression of activities that includes a general sequence of events, albeit a sequence open to alternative paths and tangents. The Standards attempted to move science teachers beyond the "scientific method" that is frequently included in science textbooks and help teachers better understand the complex and sometimes non-linear ways that inquiry can be used (NRC, 1996).

As a follow up to the Standards, the National Research Council recently published a guide for teaching and learning inquiry (NRC, 2000). The document provides detailed definitions and descriptions of how inquiry can be used in the science classroom. As part of the definition of inquiry, the NRC included what they consider to be the five essential features of classroom inquiry (NRC, 2000, p. 29). The essential features are: 1) Learner engages in scientifically oriented questions, 2) Learner gives priority to evidence in responding to questions, 3) Learner formulates explanations from evidence, 4) Learner connects explanations to scientific knowledge, 5) Learner communicates and justifies explanations (NRC, 2000).

The view of inquiry presented in the National Science Education Standards is clearly a step beyond the "process view" of inquiry, but to some science educators the vision does not go far enough. For those who take a more Vygotskian perspective of education, inquiry is closely linked to bringing scientific culture into the classroom discourse (Vygotsky, 1986). According to this view, the goal is to create a classroom community where understanding is socially constructed and where students operate as a "community of scientists" creating understandings through social interactions.

While the NRC Standards clearly emphasize the role of group interactions when engaging in inquiry, the five essential features outlined above seem to lack much in the way of social connotations. In reality each child is in a socio-cultural community where his or her background and skills interacts with the class background resulting in a unique experience for each student. This can mean that the way one student experiences inquiry in the classroom may be different than another (Huffman, Lawrenz & Minger, 1997).

PART II: A MODEL FOR EVALUATING INQUIRY
IN THE SCIENCE CLASSROOM

As more and more teachers, schools and districts attempt to make inquiry an integral part of their science program, the need to evaluate its impact becomes more apparent. Because of the complex and subtle nature of inquiry, traditional methods of evaluation do not completely consider the full impact of inquiry in the science classroom. Evaluating inquiry is challenging, but it can be done effectively and efficiently if one carefully integrates multiple methods of evaluation. There are many different possible outcomes that can occur in an inquiry-based science classroom and one must cast a broad net to capture all of the possible impacts. At the same time, one must look closely at how inquiry is used in the classroom in order to understand the nuances of its implementation. Satisfying these competing demands is challenging, but can be accomplished.

One way to help balance the competing demands is to first define a model to conceptualize the concept and to help focus on the key aspects to be evaluated. The model presented in this chapter focuses specifically on inquiry; however, there are different models and perspectives of program evaluation in science education. For example, Altschuld and Kumar (1995) defined a comprehensive model for the evaluation of science education programs. In addition, other evaluators have discussed new roles and perspectives for the evaluation of science education (O'Sullivan, 1995; Stufflebeam, 2001). For the present evaluation of inquiry, it is helpful to conceptualize inquiry into three key components: abilities, procedures, and philosophy. The three different components are represented by concentric circles with abilities at the center, procedures encompassing abilities, and philosophy encompassing both procedures and abilities (See Figure 1). First, *abilities* refers to the skills that students develop as a result of inquiry in the classroom. These are the skills one would expect students to develop when they engage in inquiry-oriented science. Abilities include traditional process skills such as observing, classifying, and making predictions, and more

The three components of inquiry are each integrated into the other starting with abilities at the center of the circle and moving outwards to include procedures and philosophy. At the center lies a focus on students' process skills and abilities. The second level recognizes that inquiry is more than just skills and abilities. Abilities are part of inquiry, but inquiry also involves incorporating skills into a series of procedures or steps. Finally, at the outermost circle lies philosophy. Philosophy includes both procedures and abilities, but in addition, instruction is committed to a philosophical orientation of living though inquiry and having the whole class use inquiry as a philosophical way of operating.	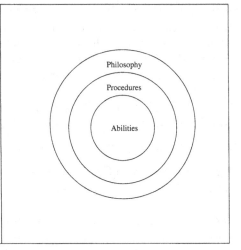

Figure 1. A Model of Science Inquiry in the Classroom.

complex inquiry skills such as designing and conducting experiments. This aspect of inquiry is characterized by a science classroom that focuses on the individual abilities and skills of inquiry. Activities tend to be isolated and focused on developing one skill at a time, such as learning how to observe or learning how to hypothesize or make predictions. These skills can be evaluated with a variety of test formats such as multiple-choice, short answer, and hands-on practical exams. Often inquiry skills are only evaluated with paper and pencil tests, but to truly gain an understanding of students' skills, hands-on tests are critical to fully evaluating the impact of inquiry.

Another key aspect of inquiry are the *procedures*. Procedures refer to the steps or sequence of activities in which students engage when they conduct inquiry. This includes the class activities an evaluator might observe, the actions students take in the classroom, the sequence of activities used by the classroom teacher, and the ways that students are engaged in the class. Using the phases of inquiry described by Champagne (2000), the process of inquiry generally occurs in four phases: precursor, planning, implementation, and closure/extension. In the precursor phase students formulate preliminary questions, analyze existing information and develop a rationale for further inquiry. In the planning phase students develop an actual plan for their inquiry that may include the design of an experiment or scientific investigation. In the implementation phase students set up their investigation, collect and analyze data, and develop conclusions from their analysis. Finally, in the closure/extension phase students communicate their findings, consider alternative ways results can be interpreted, and apply what they have learned to make a decision.

The procedures students use when they engage in inquiry are most directly evaluated through observations of the classroom; but observations can be time consuming and expensive. Because the phases of inquiry generally proceed over multiple lessons an evaluator would need to make frequent visits to the classroom. One way to ameliorate the time and expense considerations would be to use student and teacher surveys about classroom practices and activities which would serve as proxies in providing a general picture of inquiry. Using personal interviews in conjunction with surveys can help to add more information about classroom inquiry. The only caveat to this approach is that student, teacher and observers views are not always in agreement. Beam and Horvat (1975) found that science teachers viewed their classes as more inquiry oriented than did students or external observers. In addition, Fraser (1994) found that student perceptions of the classroom learning environment are more strongly related to student outcomes than teacher perceptions.

The third aspect of inquiry is *philosophy*. Philosophy is an essential feature of inquiry because it provides the rationale for the actions in a classroom. As an evaluator, one may observe students designing an experiment, but without an understanding of the teachers' philosophical perspective, it is difficult to fully judge the activity. For example, a teacher may have students do a laboratory activity, but it is one thing to simply do a series of lab activities, and quite another to do the lab in the context of a classroom where students are engaged as a community of scholars. For some science teachers, the classroom "lives" through inquiry where inquiry is a guiding philosophical perspective used to organize the science classroom and to encourage a discourse designed to construct understanding.

The philosophy of the classroom can be evaluated to some extent through surveys and observations. On the other hand, truly understanding the philosophical perspective of the teacher and the classroom requires more in-depth qualitative methods. A case study approach that includes interviews and observations can help to provide the details needed to evaluate the complex nature of the teacher's philosophy and the way that that philosophy is carried out in the classroom. But, as anyone who has conducted case studies knows, they can be quite costly and intensive in terms of the demand on the evaluator's time. Despite their value in better understanding the experiences of participants, in most program evaluations there are not enough resources to include sufficient cases to represent the whole range of experiences.

PART III: AN EXAMPLE FROM THE FIELD: EVALUATING AN INQUIRY-BASED SCIENCE PROGRAM

I recently co-directed a national evaluation of a new inquiry-based science program that was designed and implemented by the National Science Teachers Association (NSTA).[1] The purpose of the project was to help high school teachers implement inquiry-based science as described in the NRC Science Standards. The new curriculum was designed to help teachers meet both the 9–12th grade content standards and the science as inquiry standards. Activities were developed to assist students in developing the abilities necessary to do scientific inquiry. The new curriculum materials were initially used in 13 high schools across the U.S. in over 100 science classrooms. The basic tenets underlying the project were: every student should study the four science subjects of biology, chemistry, physics and the earth/space sciences every year; science teaching should take into account students' prior knowledge and experience; students should be provided with a sequence of content from concrete experiences and descriptive expression to abstract symbolism and quantitative expression; students should be provided with concrete experiences with science phenomena before the use of terminology that describes or represents those phenomena; concepts, principles, and theories should be revisited at successively higher levels of abstraction; teaching should utilize the motivational power of relevance by connecting the science learned to subject areas outside of science; coverage of topics should be greatly reduced with an increased emphasis on greater depth of understanding of fewer fundamental topics.

Evaluation Methods

The purpose of the evaluation was to examine the impact of the new inquiry science program on students, teachers and the science classroom. Because of the complex nature of an inquiry science program, a variety of methods were necessary to fully capture the impact of the new program. Whenever an evaluator examines a complex program, choices must be made to focus on the key aspects of inquiry, while at the same time attempting to evaluate all the complexities in a cost efficient manner.

A mixed-method evaluation approach was used in an attempt to balance these competing demands and to integrate both quantitative and qualitative approaches. The evaluation used an integrated design where

---

[1] This evaluation was supported by a grant from The National Science Foundation; award number 9714189. PI Frances Lawrenz, co-PI Douglas Huffman.

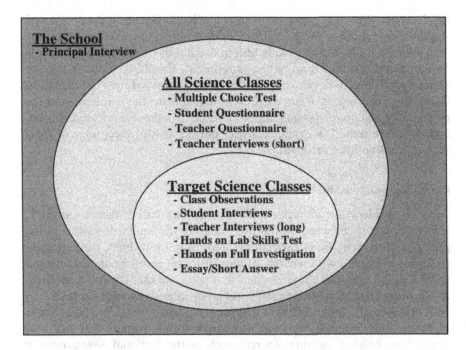

Figure 2. Sources of Information for the Inquiry Evaluation.

case studies of classrooms were integrated into the larger quantitative study (Greene & Caracelli, 1997). The evaluation included a mix of quantitative data from tests and questionnaires along with qualitative data collected through interviews, observations and case studies. The evaluation purposely included a mix of qualitative and quantitative data in order to examine the program from multiple perspectives (Yin, 1994). The evaluation was also comparative in nature in that students in the existing science classes were compared with students in the subsequent year that used the new NSTA inquiry science approach.

In this evaluation, a nested data collection design was used (See Figure 2). In all 13 schools quantitative data were collected in all 9th grade science classes. In these science classes, we administered student and teacher surveys, briefly interviewed teachers about their class activities, and administered multiple-choice science achievement tests. Within each school, three science classes were randomly selected for more in-depth data collection. The targeting of a smaller number of classes for in-depth data collection allowed us to economically collect richer information in a select number of classes, but still collect information in all 13 schools. In each of these target science classes, we observed science classes and interviewed students and teachers. We then also administered an open-response science test, a hands-on laboratory skills test, and a hands-on test where students designed and conducted an entire experiment. Finally, to better track the effects over time, in-depth case studies were integrated into the design to follow the experiences of three different teachers as they implemented the new inquiry science program. The three teachers were interviewed approximately every month over the two-year period about perceptions of their science class and the program.

Evaluating Student Abilities

To evaluate the impact of this new inquiry-based science program, information was collected on the three aspects of inquiry described in the previous section: abilities, procedures, and philosophy. First, student abilities were evaluated with a series of four different test formats that ranged from a broad multiple choice format that could be quickly and cost effectively administered, to more costly and labor intensive hands-on tests.

Specifically, four different tests of students' inquiry ability were included. The broadest test was a multiple-choice format that included items from existing national sources such as the National Assessment of Educational Progress (NAEP), International Assessment of Educational Progress (IAEP), and the Second International Science Study (SISS). The test included both content and inquiry items matched to the NRC Science Standards. A panel of science educators selected items to match the science standards and external science educators reviewed the selection of items to confirm that they did indeed match the standards. Since the test was rather long, matrix sampling was used where half the students in each classroom took one form of the test and half took the other form so each student only answered half the items. The multiple-choice inquiry items on this test included questions from the above sources such as providing students a research question and asking them to choose the best experimental design. See the example item in Figure 3. There are many multiple-choice items such as these that have already been developed by NAEP and others. These items have been extensively field-tested and are readily available to evaluators. They provide a cost effective and reliable way to measure inquiry skills. Although some science educators believe that inquiry cannot be properly measured with multiple-choice items, in this evaluation we found significant correlations between multiple-choice items and more authentic hands-on items. In addition, we found that hands-on lab skills testing cost approximately 17 times more than multiple-choice testing (Lawrenz, Huffman, & Welch, 2000).

Which of the following procedures is essential in an experimental study designed to investigate the effects of vitamin K in the diets of humans?

A. Make sure that all the study subjects get the same amount of vitamin K.

B. Use only students as study subjects.

C. Use several different brands of vitamin K.

D. Make sure that all study subjects are kept in different environments.

E. Divide the study subjects into experimental and control groups.

Figure 3. Multiple Choice Inquiry Item.

The second format that was used to evaluate inquiry abilities was an open-response format. These items asked students to plan an investigation or draw conclusions from data given to them. Each item usually focused on just one aspect of inquiry such as generating research questions, planning or interpreting information. In the item shown in Figure 4, students were asked to describe factors that need to be considered in calculating the financial and environmental costs of genetically engineered bacteria, and come up with possible research questions. The open-response questions were also selected from the previously mentioned existing national tests to match the NRC standards. The items were included on the same test as the multiple-choice items and were therefore split across the two forms of the test so each student only answered six items. Open-ended items are more expensive and more time-consuming to administer than multiple-choice items, but

Some people want to use genetically engineered bacteria to prevent frost damage to valuable plants such as strawberries and potatoes. When these bacteria are sprayed on leaves they lower the temperature at which frost forms.

Statement: People who oppose the use of genetically engineered bacteria assert the bacteria are not ecologically safe.

Describe the *factors* that need to be considered in calculating the financial and environmental costs of using genetically engineered bacteria to prevent frost damage. What *questions* would you ask to gather data to support or refute the statement above? For example, you might ask whether bacteria engineered for frost prevention cause disease in plants or animals.

Figure 4. Open-Response Inquiry Item.

they do provide more detailed information about students' inquiry abilities. The item in figure 4 measures students' ability to construct their own research questions; a performance-based skill that cannot be easily measured with a multiple-choice test format.

The hands-on lab skills test we used in this evaluation was designed to measure the process skills expected in an inquiry classroom. The test included five different laboratory stations, one for each content area (earth science, life science, physics, and chemistry) and one on the use of science instruments. A sample lab skills station is depicted in Figure 5. The laboratory stations used in this test were modeled after performance tests from the National Assessment of Educational Progress (NAEP), Second International Science Study (SISS), and the International Assessment of Educational Progress (IAEP). The stations were set up in one room, and students moved from one to the next until they had completed all five stations. In order to administer this test in a cost efficient manner a random sample of ten students in each of the target science classes took this test.

STATION TWO: CHEMICAL INDICATORS

The test strip is an indicator that turns from yellow to green in the presence of a certain type of sugar. Iodine is used as an indicator to test for starch. It will turn a starch solution blue-black in color. Before you are three cups labeled A, B, and C. One contains a sugar solution, another a starch solution, and a third contains neither starch nor sugar.

You are to determine the contents of each cup.

1. Using the information above, what will you do to determine which cup contains starch and which contains sugar? Write out your plan then *CARRY OUT THE EXPERIMENT.*

2. Record your observations.

3. On the basis of your observations, answer the following questions.

A. Which sample contains sugar? _____

   What are your reasons for this conclusion? _____

B. Which sample contains starch? _____

   What are your reasons for this conclusion? _____

Figure 5. Hands-on Lab Skills Station.

Finally, the most authentic test format used to evaluate inquiry was a hands-on test that required students to design and conduct an entire experiment. The test was designed to measure students' ability to design, conduct and draw conclusions from a scientific experiment. This test was the most authentic in that it most closely matched the entire process of inquiry, however, it was expensive and time consuming to administer because each student was given 35 minutes to complete the entire task. It was modeled after the test developed by the Assessment of Performance Unit (APU) (Welford, Harlen, & Schofield, 1985). In this test, students were given approximately 25 live sowbugs, and asked to conduct an investigation to determine which of four environments the sowbugs preferred; damp/dark, dry/dark, damp/light, or dry/light (see Figure 6). Again, because of the amount of time and expense involved in taking this test, two students in each target class were randomly selected to complete the task.

Overall, the four different test formats we used to evaluate inquiry (multiple-choice, open-response, hands-on lab skills, and hands-on full investigation) provided a range of data on students' inquiry abilities. It is especially advantageous to include both written items, which are relatively easy to administer and score, along with a sample of hands-on items. While the latter are expensive and time consuming, an evaluator could use matrix sampling techniques to efficiently gather information on students' actual performance when faced with an inquiry task.

Evaluating Inquiry Procedures

Evaluating inquiry procedures also requires multiple data collection methods because there are so many different aspects to inquiry and ways to use it in a science classroom. In the evaluation of the NSTA inquiry science we used three different techniques: class observations, interviews, and written surveys. We triangulated the data collection by gathering information from the evaluators' perspective through class observation, the teachers' perspective through interviews and surveys, and the students' perspective through interviews and surveys. The questions on the interviews and surveys used in this evaluation were matched so that items focused on the same aspects of the classroom. The survey provided broad general descriptions of class activities, while the follow-up interviews provided more detailed explanations. The survey used in this study was developed by the author to assess students' and teachers' perceptions of their science classroom learning environment (Huffman, Lawrenz, & Minger, 1997). The survey included six different scales: Relevancy, Critical Voice, Difficulty, Involvement, Sequence, and Experimental Design. The Relevancy scale measures the extent to which science was relevant to students'

ISOPODS

Isopods are small bug-like creatures with many legs. Two students were carrying out a survey of isopods. They were arguing about the sort of place in which isopods would prefer to live. You are to plan an investigation to settle the argument. It is no good asking isopods what kind of place they like to be in—you have to find out by watching where they go. This is what you have to find out:

If isopods are given a choice of the four places below, which one do they choose to live in? A place which is:

DAMP and DARK

DRY and DARK

DAMP and LIGHT

DRY and LIGHT

You can use any of the following materials:

| Water | Scissors | Black Tag board | Stop watch | Black plastic |
| Scissors | Ruler | 1/2 inch thick foam | Spoon | 25 isopods |

1. Draw a picture of how you set-up the investigation.

2. Explain in words what you did to conduct the investigation

3. Explain in words the results of your Investigation.

Figure 6. Hands-On Full Investigation.

out-of-school lives, while the Critical Voice scale measures the extent to which students believe it is acceptable and beneficial to question the teacher's pedagogical plans and methods (Fraser, 1994). The Difficulty scale measures the students' perception of the academic difficulty of their class-room. The Involvement scale assesses students' perceived personal involvement in the classroom. The Sequence scale measures the extent to which the sequence of activities used by the teacher was inquiry-oriented, for example, whether or not the teacher used hands-on activities before lec-turing to the class. The Experimental Design scale was developed to measure the students' perception of the degree of open-endedness in the design of the experiments that were conducted in the classroom. All items were written in the personal form as recommended by Fraser and Tobin (1991), in order to more accurately elicit students' personal perceptions of the classroom learning environment. Figure 7 contains examples of items from each of the scales as well as their origins.

The student questionnaire also included a long list of items about the frequency of different class activities. Included were questions on how often teachers used lab activities, had students form hypotheses, analyze data, and examine alternative explanations. The purpose of these items was to

Example Items	Origin of Item
Difficulty I find the work hard to do.	Learning Environment Inventory (LEI) (Fraser, Anderson, & Walberg, 1982)
Relevance I learn interesting things about the world outside of school.	Constructivist Learning Environment Survey (CLES) (Taylor, Fraser, & White, 1994)
Critical Voice I feel unable to complain about anything.	Constructivist Learning Environment Survey (CLES) (Taylor, Fraser, & White, 1994)
Involvement I am included in class discussions.	Classroom Environment Scale (CES) (Moos & Trickett, 1987)
Sequence We begin new topics with a lab activity.	Science Classroom Inventory (SCI) (Huffman, Lawrenz, & Minger, 1997)
Experimental Design I am allowed to go beyond the regular lab activity and do some experimenting of my own.	Modified SLEI (Fraser, Giddings & McRobbie, 1992)

Figure 7. Example Survey Items and their Origins.

describe the extent to which teachers used inquiry methods in their class. These classroom activity items were factor-analyzed to create three scales: inquiry, mixed and traditional teaching. The inquiry factor contained five items mostly related to students examining data or experiments; the mixed factor contained four items focused on working and sharing in groups; and the traditional factor contained six items about students listening to the teacher talk or read. Using surveys to evaluate teaching processes like inquiry is probably one of the most common and cost effective means of evaluation. While such use is common, well designed surveys with reliable and valid scales are rare with one of the major criticisms being that they do not accurately reflect the true nature of the classroom. This may be true for poorly designed surveys that have not been carefully constructed and pilot-tested. The process of designing, pilot-testing and forming reliable scales using techniques like factor analysis is a major undertaking, but when done carefully and thoughtfully it can produce a survey instrument that reliably represents classroom life. Follow-up interviews with students can help to verify the validity of information collected through surveys. One could also use focus groups to collect interview data from a wider range of students rather than just one-on-one interviews.

Besides the surveys, classroom observations were conducted to document activities in the actual classroom settings. The observation protocol for this evaluation was developed from the descriptions of ideal science classrooms in the NRC Standards (1996) (See Figure 8). The observations were of teaching practices and aspects of the learning environment that were included on the student questionnaire. The main purpose was to evaluate the extent to which inquiry teaching techniques were used in the classroom. During the class period, the observer recorded class activities every five minutes. Class profiles were created from the observation data by calculating the average percent of class time spent on various activities such as laboratory activities, lectures, discussion, etc. The activities were then categorized as either inquiry, traditional, or a combination of inquiry and traditional teaching. This allowed us to characterize the extent to which each class was using inquiry teaching methods. When lab activities were observed, the level of open-endedness was categorized using Schwab's definitions as previously described (Schwab, 1960). The main disadvantages of observations is that they are quite time consuming and they do not tell one the motivation behind an action. Although types of class activities can be observed and classified, follow-up interviews are needed to fully understand the teachers' motivation and rationale for using a particular activity or teaching technique.

Follow-up interviews with teachers, along with a written survey, provided further information about how often teachers used various

1.	2.	3.	4.	5.	6.	7.	8.	9.	10.	11.	12.
0–5	5–10	10–15	15–20	20–25	25–30	30–35	35–40	40–45	45–50	50–55	55–60

Activity Checklist:

Please fill in the approximate time for the instruction strategy used in this class:

Type of Instruction		*Type of Instruction*	
L	lecture	TSW	text seat work
LWD	lecture with discussion	RSW	other reading seat work
CD	class discussion	LWW	lab write up
HOA	hands-on activity/materials	WW	worksheet work
SGD	small group discussion	D	demonstration
AD	Administrative tasks	CL	coop learning (roles)
		TIS	teacher interacting w student

other_____

13. **Description of Setting:** (seating/size/facilities/lighting/ethnic/historic)

14. **Student involvement:**   Almost Always   Sometimes   Almost Never
    Description:

15. **Critical Voice:**   Positive evidence   Negative evidence   No evidence
    Description/Example:

16. **Instruction:**   Traditional   Mixed   Inquiry
    Description:

17. **Relevancy:**   Almost Always   Sometimes   Almost Never
    Description: (useful in real-life/made real-world connection)

18. **Lab Activity—Schwab's Levels:**
    1. Labs define the problem to study and include detailed instructions on how to conduct lab.
    2. Labs define the problem to study and include *some* suggestions on how to conduct the lab, but students can also use some of their own procedures.
    3. Labs define the problem to study, but students *design their own procedures* to conduct the lab.
    4. Students come up with both their *own problems* to study as well as their *own lab procedures.*

19. **Description of Lab Activity:**

20. Overall, In what ways did the teacher encourage inquiry?

Figure 8. Observation Schedule for Inquiry Evaluation.

instructional techniques, and the teachers' perceptions of the learning environment of their classrooms. Many of the items on the student and teacher survey were identical so that comparisons could be made in student and teacher perceptions. The interviews and surveys for teachers were critical to helping to interpret observations and better understand the motivation behind the activities that were observed.

Evaluating Philosophy

Determining the teachers' philosophy of inquiry is one of the most challenging aspects of evaluating inquiry. Written surveys as described above can certainly be used to gather general information about teachers' philosophical points of view, but more detailed information is usually required to fully evaluate teachers' philosophies. In the evaluation of the NSTA inquiry science, case studies were used to closely follow the experience of three different teachers as they transformed their curriculum from a traditional lecture-based approach to a more inquiry-based approach. They were interviewed approximately every month over the two-year period about their perceptions of their science class and their change efforts. The interviews and visits to these teachers' classrooms looked at their instruction, the curriculum, and their philosophy of teaching and learning. Themes emerged in each case, and then the three cases were subjected to a cross-case analysis to build abstractions across them (Merriam, 1998). The case studies were invaluable in helping to better understand the experience of the teachers. The on-going conversations and updates on progress provided rich data about the process of implementing inquiry-based science. The surveys and interviews with students described previously provide a snapshot of the classroom, but to really understand the on-going challenges in the classroom, the case studies proved to be the perfect method.

Overall, the results of this evaluation indicated that the teachers significantly altered the learning environment of the classroom through the use of the NSTA inquiry science curriculum. Twelve of the thirteen schools exhibited significant changes in the frequency of teachers using inquiry methods such as having students do experiments, formulate hypotheses, and analyze data. These changes were found on student and teacher questionnaires, and were documented through evaluators' observations of classes. However, despite the changes in the learning environment, altering student achievement was found to be much more challenging. Less than half the schools were able to outperform comparison classes. In addition, even the schools that had higher achievement in treatment classes were unable to sustain the achievement gains for more than several years. Furthermore, the schools were also unable to maintain the frequent use of inquiry teaching methods. By the third year of the program the frequency of inquiry teaching methods had declined in most schools. On the other hand, the science teachers indicated that they were still committed to inquiry science, suggesting that a philosophical shift in their beliefs has remained as a result of the project. Overall, the results of this evaluation suggest a national curriculum development project can help teachers create an inquiry-oriented learning environment; while at the same time, long-term reorganizing that sustains changes is more challenging.

PART IV: IMPLICATIONS FOR THE FUTURE

As views of inquiry become more complex and more diverse, so must the methods used by evaluators. The NRC Science Standards provide clear definitions of what is meant by inquiry and how to use inquiry to teach science (NRC, 2000). However, views of science always change and views of inquiry will likewise continue to evolve. There are science educators who take a more social constructivist view of inquiry than the view presented in the NRC Science Standards. Duit and Treagust (1998) recently described a variety of perspectives on learning in science including social-constructivist, phenomenological, and anthropological. Traditionally, knowledge has been viewed as something possessed by an individual. New perspectives on learning view knowledge as something between an individual and society, and place more emphasis on the importance of learning in a social milieu and the cultural implications of learning. The notion of socially constructed knowledge has implications for inquiry and for what it means to learn in an inquiry classroom. It is no longer enough for students to merely engage in inquiry on their own. Students are now expected to participate in learning communities and engage in the process of socially constructing knowledge.

These new perspectives on learning have implications for science education evaluation. As science education moves beyond the NRC Science Standards and begins to focus on different ways of conceptualizing science education and what it means to learn science, the field of evaluation will need to consider alternative methods as a means of better understanding the science classroom. One of the most promising approaches is mixed-method evaluation. Neither quantitative nor qualitative approaches alone provide the information needed to evaluate a complex concept like inquiry. Mixing evaluation methods provides the flexibility needed to match the method with the variety of outcomes that can result in an inquiry-oriented science classroom. There are practical guidelines provided by the National

Science Foundation on how to employ mixed-method evaluations (NSF, 1997). The example presented in this chapter was a mixed-method integrated design, where qualitative methods were integrated into a predominately quantitative evaluation. As described by Greene and Caracelli (1997), methods can mixed in many different ways. There are component designs where qualitative and quantitative methods remain distinct aspects of the evaluation, and there are integrated designs where qualitative and quantitative methods are combined in unique ways. Within the integrated framework there are multiple ways to integrate methods, such as iteratively, in a nested manner, or holistically. Even though a true purist would claim that methods cannot be mixed because they represent different philosophical paradigms, I believe that the practical demands of evaluation require one to mix methods to truly understand most programs.

New definitions of inquiry, new methods of evaluation, new ways of mixing methods; all of these developments make the field of science education evaluation quite exciting. As the field of science education continues to advocate for making inquiry a central part of every science classroom, there is hope that one day science instruction as we know it today will change. The same is true for the field of evaluation. There is hope that one day we will see a more participatory approach to evaluation (Patton, 1997; Cousins & Earl, 1995). Instead of evaluation being viewed as an external activity done by outsiders who visit a school to "evaluate" a science program, there is hope that the philosophy of evaluation will be internalized by teachers, students, and administrators so that all the stakeholders will become actively involved in the process of evaluation. There is even hope that everyone involved will engage in "inquiry" both inside and outside the classroom so that we can all gather and analyze the data needed to improve the quality of science instruction for all students.

REFERENCES

Altschuld, J. W., and Kumar, D. (1995). Program evaluation in science education: The model perspective. *New Directions for Program Evaluation*, 65:5–17.

Beam, K. J., and Horvat, R. E. (1975). Differences among teachers' and students' perceptions of science classroom behaviors and actual classroom behaviors. *Science Education*, 59:333–344.

Bruner, J. (1960). *The Process of Education*. New York: Vintage.

Champagne, A. B., Kouba, V. L., and Hurley, M. (2000). Assessing inquiry. In Minstrell, J. and van Zee, E. H. (Eds.), *Inquiring into Inquiry Learning and Teaching in Science* (pp. 447–470). Washington, DC: AAAS.

Cousins, J. B., and Earl, L. M. (Eds.). (1995). *Participatory Evaluation in Education: Studies in Evaluation Use and Organizational Learning*. London: Falmer.

DeBoer, G. E. (1991). *A History of Ideas in Science Education: Implications for Practice*. New York: Teachers College Press.

Duit, R., and Treagust, D. F. (1998). Learning in science—From behaviourism towards social constructivism and beyond. In Fraser, B. J. and Tobin, K. G. (Eds.), *International Handbook of Science Education* (pp. 3–25). The Netherlands: Kluwer.

Dewey, J. (1910). Science as subject matter and as method. *Science*, 121–127.

Dewey, J. (1938). *Logic: The Theory of Inquiry*. New York: MacMillan.

Fraser, B. (1994). Research on classroom and school climate. In Gabel, D. (Ed.), *Handbook of Research on Science Teaching and Learning* (pp. 493–541). New York: Macmillan.

Fraser, B. J., Anderson, G. J., and Walberg, H. J. (1982). *Assessment of Learning Environments: Manual for Learning Environment Inventory (LEI) and My Class Inventory (MCI).* (Third Version). Perth: Western Australian Institute of Technology.

Fraser, B. J., Giddings, G. J., and McRobbie, C. J. (1992). Assessing the climate of science laboratory classes. *What Research Says to the Science and Mathematics Teacher No. 8.* Perth: Key Centre for School Science and Mathematics, Curtin University of Technology.

Fraser, B. J., and Tobin, K. (1991). Combining qualitative and quantitative methods in classroom environment research. In Fraser, B. J. and Walberg, H. J. (Eds.), *Educational Environments: Evaluation, Antecedents, Consequences.* (pp. 271–292). London: Pergamon.

Greene, J., and Caracelli, V. (Eds.). (1997). Advances in mixed-method evaluation: The challenges and benefits of integrating diverse paradigms. *New Directions for Evaluation*, 74.

Greene, J., and Caracelli, V. (1997). Crafting mixed-method evaluation designs. *New Directions for Evaluation*, 74:19–32.

Harms, N., and Kohl, S. (1980). *Project Synthesis*. Final report submitted to the National Science Foundation. Boulder, CO: University of Colorado.

Huffman, D., Lawrenz, F., and Minger, M. (1997). Within-class analysis of ninth-grade science students' perceptions of the learning environment. *Journal of Research in Science Teaching*, 34(8):791–804.

Lawrenz, F., Huffman, D., and Welch, W. (2000). Policy considerations based on a cost analysis of alternative test formats in large scale assessments. *Journal of Research in Science Teaching*, 37(6):615–626.

Merriam, S. B. (1998). *Qualitative Research and Case Study Applications in Education.* San Francisco: Jossey-Bass.

Moos, R. H., and Trickett, E. J. (1987). *Classroom Environment Scale Manual* (2nd ed.) Palo Alto, CA: Consulting Psychologists Press.

National Educational Association. (1893). *Report of the Committee on Secondary School Studies.* Washington, DC: U.S. Government Printing Office.

National Research Council. (1996). *National Science Education Standards.* Washington, DC: National Academy Press.

National Research Council. (2000). *Inquiry and the National Science Education Standards: A Guide for Teaching and Learning.* Washington, DC: National Academy Press.

National Science Foundation. (1997). *User-Friendly Handbook for Mixed Method Evaluation.* In Frechtling, J. and Sharp, L. (Eds.), Washington, DC: National Science Foundation.

O'Sullivan, R. G. (Ed.). (1995). Emerging roles of evaluation in science education reform. *New Directions for Program Evaluation*, 65, spring, 1995.

Patton, M. Q. (1997). *Utilization-Focused Evaluation.* Thousand Oaks, CA: Sage.

Schwab, J. J. (1958). The teaching of science as inquiry. *Bulletin of the Atomic Scientists*, 14:374–379.

Schwab, J. J. (1960). Enquiry, the science teacher, and the educator. *The Science Teacher*, 6–11.

Shymansky, J., Kyle, W., and Alport, J. (1982). Research synthesis on the science curriculum project of the sixties. *Educational Leadership*, October, 1982, 63–66.

Smith, A., and Hall, E. (1902). *The Teaching of Chemistry and Physics in Secondary School.* New York: Longmans, Green.

Stufflebeam, D. L. (2001). Evaluation models. *New Directions for Evaluation*, 89, spring, 2001.

Taylor, C. S., Fraser, B. J., and White, L. R. (1994). *CLES: An Instrument for Monitoring the Development of Constructivist Learning Environments.* Paper presented at the annual meeting of the American Educational Research Association, New Orleans, April 1994.

Vygotsky, L. S. (1986). *Thought and Language.* Cambridge, MA: MIT Press.

Welch, W. W., Klopfer, L. E., Aikenhead, G. S., and Robinson, J. T. (1981). The role of inquiry in science education: Analysis and recommendations. *Science Education*, 65:33–50.

Welford, G., Harlen, W., and Schofield, B. (1985). Practical testing at ages 11, 13, and 15. *APU Science Report for Teachers: 6.*

Yin, R. (1994). *Case Study Research: Design and Methods.* Thousand Oaks, CA: Sage.

CHAPTER 10

Distance Learning in Science Education
Practices and Evaluation

John R. Cannon

INTRODUCTION

To voice that there currently exists an explosion in distance education would be a vast understatement of enormous proportion. It was estimated in 2000 that approximately 17,000 on-line courses were offered on the Internet (Charp, 2000). In describing the effect of the Internet and the World Wide Web (WWW) on education, Sun Microsystems co-founder John Gage reports, "it's impossible to overstate its impact . . . the Web is real, it's not diluted down into a text book and five years out of date" (Curriculum Administrator, 2000). Murray Goldberg, president of WebCT™, one of the predominate software companies for offering courses on the web, supports Gage's claim. Goldberg states, "In the first 16 months WebCT has been in business (to February 1, 1999), we have sold over 1,200 licenses to institutions in over 33 countries. In the newest version of the software, the company

John R. Cannon, Department of Curriculum and Instruction, College of Education, University of Nevada, Reno, MS 282, Reno, NV 89557.

Evaluation of Science and Technology Education at the Dawn of a New Millennium, edited by James W. Altschuld and David D. Kumar, Kluwer Academic / Plenum Publishers, New York, 2002.

claims, "Now setting up your on-line course is so easy, you can do it in 15 minutes. That's the beauty of WebCT 3.0" (http://www.webct.com/).

Pennsylvania State University, in its Distance Education Catalog, 1999–2000, states that it "has been a pioneer in distance education since 1892," founding one of the nation's first correspondence study programs (The Pennsylvania State University, 1999, p. 1). With this new catalog also comes a new title for Penn's Distance Education Program—The Penn State World Campus. No longer does one need to physically attend courses at the Penn State campus to prepare for several of their associate degrees, graduate in any of the twelve certificate programs (e.g., Counselor Education: Chemical Dependency), or even complete a Master's degree in Elementary Education.

Institutions of higher education are flocking toward catering to virtual students of the distance education student body (Fraser, 1999; Hara and Kling, 2000; The Institute for Higher Education Policy, 1999). This comes as no surprise as in many colleges and universities, projected student enrollments are down. Edelson (1998) reported that since 1995, courses taught via the Web had grown into the thousands, and in 1996, estimates of students enrolled in such courses topped one million. During the year 2000, projections suggest enrollments to be nearing three million. The resulting increase of three million additional tuition paying students to colleges and universities worldwide requires no new expensive classrooms to sit in or, more importantly for some campuses, no additional parking spaces for their automobiles.

The National Center for Education Statistics disclosed that in 1995, "one-third of higher education institutions offered distance education courses" and "another 25 percent indicated plans to begin courses within 3 years" (National Center for Education Statistics, 1997, p. 1). "Internet courses," Edelson (1998) maintains, "have clearly emerged as the technology-of-choice" for part-time adult students who cannot attend orthodox college courses (p. 1).

The University of Phoenix is clearly one of the nation's leaders in offering such an alternative means for completing a college degree. The University's web site reports that it "is now one of the nation's largest private accredited universities . . . provid[ing] a relevant, real-world education to working adults at more than 92 campuses and learning centers in the US, Puerto Rico and Canada, and via the Internet" (http://www.phoenix.edu/faculty/index.html). The web site also posts that enrollment comprises over 75,000 degree-seeking students, of which more than 12,000 students are taking advantage of degree programs offered via the Internet through University of Phoenix On-line—one of the United State's leading providers of on-line education with more than

93,000 students earning their degree from University of Phoenix. (http://www.phoenix.edu/faculty/index.html).

Many questions can, and should, be raised regarding how this manifestation of on-line education programs will impact traditional institutions of higher education and their faculty, staff, and physical plant employees in the not-so-far future. Will the increased demand for on-line courses change the definition and role of faculty, both those associated with distance education and those not? How might these changing roles impact the often already stretched faculty workloads at colleges and universities? Will there ever be a significant shift of enrollments away from the traditional land grant, locale-specific universities to flourishing Internet-based institutions of higher education? Will faculty ultimately become freelance instructors, of a sort, not specifically contracted with any college or university full time?

The answers to such questions are not easily understood in this rapidly evolving climate of distance education. The increasing momentum, however, associated with this movement is unmistakable. Whether or not one agrees with the validity of offering academic credit through distance education programs, one fact remains crystal clear—distance education programs will become more attractive and increasingly available to students of all ages, genders, and cultural backgrounds in the future.

The purpose of this chapter is to review any special concerns that may emerge when considering content science and science education courses made available on-line via the Internet. Some of these include student access to the Internet, technological savvy, hardware and software requirements needed for success, evaluation of student achievement in distance education programs (especially during any practical laboratory of field-based work), and pedagogical restraints. Most importantly, best practice in educational endeavors must be research based. This movement toward offering more distance education and on-line versions of both new and traditional courses must stand the scrutiny of the research community. At the present time, the literature base supporting the movement of distance education is a collection of reports applying typical traditional measures, such as multiple-choice content exams, to, what some might claim, very atypical teaching methodologies and venues. Clearly, much more empirical work is needed upon the distance education phenomena within the science education community in order to better serve students, faculty, and institutions of higher learning of all types.

Content science and science education professors, perhaps, may have differing concerns, if any, relating to the list previously mentioned. It may also be found, when considering distance education courses and programs for science majors or pre-service science teachers, that the technology favors one over the other.

REVIEW OF THE LITERATURE

During this time of redefining the term "distance education" (synonymous with distance learning), readers may forget that the concept of learning content at one's own pace, any time and any place, is not as revolutionary as one might believe (The Institute for Higher Education Policy, 1999). Sherry's 1996 work *Issues in Distance Learning* reported that "distance learning has existed for well over 100 years. Correspondence courses in Europe were the earliest form of distance learning, and correspondence study remained the norm for distance learning until the middle of [the twentieth century] when instructional radio and television became popular" (cited in Imel, 1998, p. 1).

Even within this writer's lifetime, there are recollections of local telecourses, produced by a small Catholic college in Montana during the early 1960's, broadcasting educational programming early in the morning or late at night, typically on weekends. All broadcasts were in black and white, long before color television became a standard. Now, approximately forty years later, a new standard is emerging. This emergence is moving the distance learning ranks, typically huddled around black and white televisions often found only in the living rooms of the past, to more comfortable workstations within a home office. Dazzling colors, animated graphics and demonstrations, and often accompanying video transmissions, either in real time or downloaded to a personal computer, are part and parcel of the redefinition of distance education in contemporary society.

This review of the literature will be slightly different from typical reviews that largely report the thoughts, comments, and results of other authors' works published in print media. The bulk of the literature, which examines distance education, is mostly found posted on the medium that carries it to millions of students, namely the World Wide Web, or the Internet. While many Internet users are skeptical of information posted on the Web, and often rightly so, it still remains the largest repository of information describing distance education, along with its successes and failures. As Gage, of Sun Microsystems stated, the Web is "the morning newspaper from ten thousand newspapers around the world, it's the Encyclopedia Brittanica, the Japanese Art Museum, the British Museum, it's Beijing University and it's all free on the [students'] screens (Curriculum Administrator, 2000, p. 4).

Many organizations, companies, and institutions of higher education support web sites that share information regarding their own distance education program or that of others. Therefore, while it may seem that this review might rely too heavily upon the electronic literature available, it is this writer's intention to offer readers the most timely information regarding the evaluation of distance education as it relates to science content teaching and science education. As with any review of the literature, the final judgment rests with the reader.

Definition of Distance Education and Its Relationship to Science and Teaching

Imel (1998) defines distance learning as situations where "teachers and learners are separated by time and distance" (p. 1). The Institute for Higher Education Policy (TIHEP) (1999) defines distance education as:

> "synchronous communication," which occurs when teacher and student are present at the same time during instruction—even if they are in two different places—and "asynchronous communication," which occurs when teachers and students do not have person-to-person direct interaction at the same time or place. (p. 11)

TIHEP (1999) also notes that distance education is defined by certain key concepts, including ". . . a combination of media . . . [including] television, video tapes, audio tapes, video conferencing, audio conferencing, e-mail, telephone, fax, the Internet, computer software, and print" (p. 11).

TIHEP (1999) reports, "It should be emphasized that the review [of distance education studies] provided striking evidence of the fact that there is a relative paucity of true, original research dedicated to explaining or predicting phenomena related to distance learning" (p. 13). This point is greatly amplified when investigating the effectiveness of science teaching via distance education, although some case studies of college-level science courses (undergraduate and graduate) and high school science courses do exist. Distance education studies specifically designed for evaluating science education courses (methods of teaching science) are non-exisistent.

The effectiveness of distance education courses is usually measured by three broad measures: student outcomes (grades and test scores), student attitudes about learning via distance education, and overall student satisfaction toward distance learning (TIHEP, 1999). Studies previously done on distance education courses, in general, suggest that distance learning courses are as effective as classroom-based instruction (Imel, 1998; Rekkedal, 1994; Russell, 1999; Saba, 1999; Schulman and Sims, 1999). It is difficult to empirically support such a claim due to the inherent flaws of the research designs, if even possible to be characterized as such, employed by the authors. These flaws include the lack of control for extraneous variables, non-randomly selected subjects, using instruments with questionable reliability and validity, and not adequately controlling for the feelings and attitudes of the students and faculty studied (TIHEP, 1999). As a result, the need for specific distance education evaluation protocols is undeniably clear.

One piece of literature that is commonly cited in support of distance learning is Thomas Russell's *The No Significant Differences Phenomenon*. This annotated bibliography cites 355 sources, dating back to 1928, claiming there are no differences in distance education versus face-to-face instructional experiences. TIHEP (1999) cautions, however, that "the overall quality of the original research [in distance education] is question-able and thereby renders many of the findings inconclusive" (p. 18).

Applied Science and Science Content Courses

Ryan (2000) outlines the success of an applied science web-based course as compared to its equivalent traditional lecture formatted course. The course was a construction equipment and methods class offered at the University of Oklahoma (OU) in 1998 and 1999. Two sections of the course, one web-based and one lecture, were available to students. The twenty-five students, from nine geographically separate universities opting for the on-line version, were required to check the web site regularly after the tradi-tional lecture version of the course where students were required to regularly attend lectures. Course content, topic sequencing, homework, and examinations were the same for both on-site and on-line classes. Ryan states, "The site [for the course] was formatted like a 'book' of organized information to be used as a class information resource" (p. 79.) Students in the on-site course took notes during the lectures. Those students in the on-line class could print out the lecture notes given to the lecture class, result-ing in class notes that were actually better in quality than hand written notes. Ryan continues with, "The trade-off, however, is that the Web par-ticipants miss the interaction determining what information is the most important. This interaction also provides a convenient format in which to answer questions" (p. 80). Many of the more common distance learning characteristics were employed in the class, including e-mail, on-line chats, and desktop video conferencing. The lecture course did not have access to the on-line course web site. The content and sequence was kept the same for each respective course. "The purpose of this parallel delivery," states Ryan, "was for direct performance assessment and administration com-parison between the two instructional strategies" (p. 80).

The Test of Logical Thinking (TOLT) was given as a pretest to both groups to determine equivalency in background, knowledge, and attitude. Both groups were deemed equivalent. Two measures were taken for pre-post course comparisons: final course grades and standard course evalua-tions required by the University of Oklahoma. Results of the pre-post comparisons revealed no significant differences in both groups on the final course grades or university course evaluations. The largest weakness, reported by the on-line group, was instructor interaction. The greatest strength, reported by the same group, was the constant availability of the course on the Web. Ryan (2000) concludes that the lack of difference in the final grades, signifies the teaching techniques and styles used in the construction methods course were suitable for both on-line and traditional lecture formats.

Collins (1997) reports on the development and implementation of a biology course offered through the Internet. The resulting course was a web-based version of one of his second-year classes "Biology 2040—Human Biology and Modern Society I." This course was predicated by a previous version designed for correspondence study. Collins points out that his interest in technology and teaching began in 1994 with some experimentation with electronic bulletin boards in trying to improve the quality of interaction between himself and his students, and student-to-student interactions. Progressing through producing graphics, which took the place of overheads in his class in 1995, he developed the first version of his course for the Web. The structure of the course is similar to many present-day web-based science content courses. The components section on the welcoming page of the course web site lists the three main course components: web-based lecture notes, book of required readings, and web conferences between students and the instructor (*http://www.det.mun.ca/dcs/courses/Biol2040.html*).

Under the "course objectives" section of the welcoming page, the first objective is listed as "To learn the basic terminology associated with human biology." There is no course component for any kind of laboratory, or practical work. Collins (1997) cites results of a comprehensive evaluation of the biology course. Ten students out of 17 (59% response rate) completed the evaluation. Six students reported that the most positive aspect of the course was being able to schedule learning time at their own convenience. E-mail access to obtain help was also highly listed. All 10 reported they would consider taking another web-based course, while only one said they would recommend the course to a friend.

Negative comments associated with the course included "No textbook; studying at the computer all the time," "time and paper wasted printing text," and "technical problems caused students to fall behind schedule early on" (p. 596). Collins (1997) concludes with, "overall the course has been successful in achieving the objectives laid down for it and is being used as a model for the development of other web-based courses . . ." (p. 596). These objectives included making the biology course more interactive, scheduling learning time at the student's convenience, becoming more familiar with technology and the Internet, and offering students more information and resource materials via the Internet.

Iowa State University also has been offering an on-line college biology course through its Project BIO program. The uniqueness of this course lies within its main delivery system of information—the RealAudio™ streaming media. This technology enables the user to see and hear information being broadcast, or "streamed" over the Internet in either a syncronous or asyncronous time frame. Ingebritsen and Flickenger (1998) state, "The practical effect is that the media content can be heard or viewed almost instantaneously . . . in either an on-demand mode or manual broadcast mode" (p. 1).

Project BIO supports the multiple learning styles of seeing and hearing of on-line lectures, doing active learning assignments, and reading on-line resources or readings from the required course textbook. The active learning assignments were designed to force the students to use the technology at hand. For example, the majority of written assignments were turned in to the instructor via the ClassNet software that was used for student/instructor interactions. Each new course topic web page included additional web sites that could be accessed for more information on that topic. Students viewed and listened to on-line lectures and carried out active learning assignments, teaching them to find and process information on the Web. During the fall of 1998, the on-lines courses for majors and non-majors included Introduction to Biology, Principles of Biology I and II, Human Physiology and Anatomy, Biology of Microorganisms, Advanced Microbiology (graduate level) and introductory economics courses. Ingebritsen and Flickenger (1998) report that the largest group of students to enroll in the on-line courses were on-campus students, citing that their choice was based upon "flexible scheduling . . . the ability for the student to work at his or her own pace," and "the novelty of taking a course via the World Wide Web" (p. 5).

In an effort to determine the effectiveness of the on-line courses, during the fall of 1997, a qualitative methodology was used to "provide a rich, thick description of this phenomenon" (Ingebritsen and Flickenger, 1998, p. 5). Three students were hand-selected to be interviewed from a web-based Zoology course that was being offered at the time. One student was chosen as being much like an "average" student in the traditional lecture Zoology 155 course, one was chosen from the adult population, and the last student was determined to be "at-risk." While the authors were clear to point out that "generalization is not a goal of qualitative research," specific themes did emerge (p. 5). The themes included positive comments regarding the availability of the instructors via e-mail, changes in attitude for the subject matter through the use of the streaming media, and enthusiasm and motivation. One student reported "This is the height of customer service" (Ingebritsen and Flickenger, 1998, p. 6). The only negative themes to emerge from the interviews were problems in using the technology.

A follow-up quantitative study compared grades and attitudes of students taking a traditional lecture section and the web-based section of Zoology 155. While no significant differences in test scores were found ($a = 0.05$), grades in the web-based course were slightly higher than the traditional course. Additionally, comprehensive final grades were higher in the web-based section than the traditional lecture section. Student attitudes were found to differ in the area of attending lectures, note taking, and retention of information. The web-based students logged-on for all lectures unlike their traditional peers, who missed an average of 6–8 classes. Web-based students also took fewer notes, using the textbook and the Internet in its place. It should be noted that students who opt for on-line courses typically demonstrate intrinsic motivation for technology and its ever growing use in course offerings. Clark's (1983) work is cited by Ingebritsen and Flickenger (1998) as a possibility for the difference in final grades. Clark maintains that the instructional design is more responsible for learning increases rather than the presentation medium. In other words, no amount of flashy technology will overcome the basic and fundamental instructional design flaws found in any course.

Lui, Walter, and Brooks (1998) investigated the effects of an on-line chemistry course offered for high school chemistry teachers in 1995. Professional development was determined to be a problem and therefore, this course, and accompanying activities, were designed to increase teacher access to chemistry teaching resources. The authors stated, "We hoped that the lack of direct contact including instruction related to hands-on work would be offset by the ease of access and the ability to try out new ideas directly with high school students in the teacher's ongoing classes" (p. 123). Small-scale laboratory activities were produced resulting in seven modules.

The module topics covered basic high school chemistry content in areas such as stoichiometry, gas laws, and mass measurements. After each module was completed, a discussion via a special chem-ed list server would follow. Each module included two distinct assignments. "Teachers are assigned one experiment to test and perform and to gather classroom data. The data from classes are shared in the course. The related teaching strategy is using a computer spreadsheet to record data from student groups" (Lui *et al.*, p. 124).

Technological requirements for teachers included access to e-mail, a CD-ROM drive (acting as an information source), the ability to accomplish

FTP's (file transfer protocols for sending and receiving files), and access to a laboratory where the small-scale activities could be performed. Small-scale chemistry kits were purchased from a commercial retailer.

The results of this experience for teachers were mixed. Mismatching of modules to local chemistry curricula proved to be problematic. Approximately half of the teachers completed the course. Those that dropped out reported it was a "low priority" within their schedules and some even went so far as to say they just wanted to see what the course was all about, fully intentioned not to see it through completion. Overall, Lui *et al.* (1998) suggest that "with one exception, the students judged the course in a very favorable light" (p. 124). Although the authors admit that during the time of this course, technologies were still considered a novelty, they do offer thoughts on future courses such as this one. The technology problems could be reduced going to a web-based format rather than strictly using a discussion listserv. Comments and results of classroom data would be posted to the course web page. Video conferencing would be included in the course at regular intervals. Finally, teachers would have to demonstrate their competence in using all technologies associated with the class.

Burke and Greenbawe (1998) report the findings of The Iowa Chemistry Education Alliance (ICEA). As stated by the authors:

> The purpose of the ICEA was to provide an opportunity for collaboration among central Iowa rural, suburban, and urban high school students and their teachers who would otherwise might have never had the occasion to interact. The teachers, in collaboration with one another and university colleagues, worked to identify chemistry concepts that served to enhance the existing high school chemistry curriculum. (p. 1308)

The objective of the program was that four chemistry teachers were to design and develop eight elementary modules to augment the traditional chemistry curriculum and "to introduce collaborative learning activities to high school chemistry students . . . linked via two-way interactive, synchronous [real-time] audio-video communication . . ." (p. 1308). "Distance education," as Burke and Greenbawe (1998) state, "promotes teamwork and collaboration rather than competition among students . . . thereby enhancing collegiality and learning. Team work and collaboration are among the objectives listed in The *National Science Education Standards*" (p. 1308). Communication technologies used in the program included:

- computer-based instruction
- computer visualizations
- web lessons
- computer-based presentations
- projects using or simulating modern instrumentation

- interactive study guides
- CD-ROMs
- laserdiscs, and
- computer simulations.

Students were involved in hands-on activities in their respective high school laboratories throughout each of the eight modules. The activities were shared over the communications network supported by the ICEA and carried via the Internet. Burke and Greenbawe (1998) report that "although the teachers covered one chapter fewer than in previous years, the students did just as well on achievement tests" (p. 1312). While only speculative, it appears this claim is based upon a comparison to the previous year's test scores. They continue that students' self-esteem was enhanced, they liked the various collaborations between learning sites, and that the topics of the modules were of interest to them as reported by the participants at the end of the program. Teachers told of students always searching for the "right" answer and that this experience helped show students that collaborations are very meaningful in analyzing data and revealing the appropriate results. The development of the evaluation tool used to base these findings upon was not discussed.

Venables (1998) offers an examination of specialized graduate-level science courses on the Internet. During the spring semester of 1996, Venables taught a *Surface Physics* course at Arizona State University. As an experiment in expanding his teaching to include web-based courses, he included students in Canada and the University of Sussex, Brighton, UK, with the 12 on-site students enrolled. Basically, Venables transformed his class lecture notes and assignments into web-compatible documents that students could access at any time. Further development of another graduate course in *Adsorption and Thin Film Growth Mechanisms* followed in 1996–97. In addition, in 1998, he taught a course in *Surfaces and Thin Films*. All of the information for these courses can be accessed through Venable's web page available at http://venables.asu.edu/grad/index.html.

The courses were based upon lecture notes, coined as "web-based" talks. The talks were linked to each other and included various web links for other resources pertaining to the subject. Venables states, "links to other laboratories and research work elsewhere can give the notes a dynamism which the printed page lacks" (p. 159). All talks can be downloaded at once for use. Students are also assigned problems to complete along with additional demonstrations and simulations that are available on a university computer server.

Venables (1998) does not suggest one type of course (on-line versus on-site) is better than the other. He does raise some fundamental issues, however, that promotes the expansion of designing science content courses for the Internet. He writes:

> . . . the web is a powerful instrument for collaboration on an asynchronous basis. Students can download material, and faculty members can interact with the student anywhere; most importantly, they don't have to be awake or concentrating at the same time, in contrast to the various forms of interactive video-based classrooms involving synchronous distance learning. If they need to talk face to face, they can arrange it electronically or otherwise to suit themselves. Moreover, students can access material put up by other groups working in related areas, and can incorporate such material into projects; this means that the student is in principle, not limited by the understanding of the local teacher. The combination of projects and resources is powerful, because projects by current students, suitably filtered, can become resources for future students. (p. 159)

He closes with a discussion about pressures from universities for faculty teaching large undergraduate courses. As a result, smaller graduate courses are offered more and more infrequently, and when offered, are no longer smaller in student enrollments. "If there is a one-line message for faculty members," according to Venables (1998), "it is to use an infrequent event to create a continuously available resource" (p. 161). The Internet is such a tool. Although not convinced about where the balance between traditional classes, textbooks, videotaped lectures, and web-based resources will be struck, Venables believes, as many in higher education, that the technology is "here to stay" and for further positive developments to occur, additional resources and structural solutions must be achieved to remain competitive with the course structures currently in place (p. 163).

Kennepohl and Last (1997) discuss a collection of on-line science courses in chemistry, biology, and physics that offer, not only the theoretical components of the content areas, but also include the practical, or laboratory, aspects of learning and experimenting. The courses are taught through Athabasca University (AU), referred to as Canada's Open University. Such courses have been available since 1973. As the authors point out:

> Of the many institutions across North America that are involved in distance education, only a few offer science courses with a substantial laboratory component. Although laboratory work is an essential ingredient in most science courses, it is difficult to incorporate such work in courses offered at a distance. For this reason, many institutions avoid offering undergraduate science courses as part of their extension programs. (pp. 36–37)

Each science course is accompanied by a course package that includes all of the instructional materials to complete theoretical requirements. Packages include textbooks, solutions to problems manuals, administrative forms, and specialized items such as computer software or audio tapes. Telephone tutors are provided to help answer student questions. Calls can be placed, toll-free, anywhere in Canada during specified hours. Students enrolled in these courses can be found potentially anywhere across Canada, assuming they have telephone and Internet access.

The rising costs of offering laboratory courses are, many times, a major concern at universities. This point is amplified when factoring in considerations for developing and implementing distance learning laboratory activities. When discussing these concerns, Kennepohl and Last (1997) write, "These challenges vary from course to course and we believe there is no single solution" (p. 37). They continue with outlining a number of such solutions that accompany their on-line science courses including "home-lab kits, residential schools, the use of laboratories in regional centers, and computer simulations" (p. 37).

The physics and organic chemistry courses at AU require a supervised laboratory session. The same is true for freshman chemistry and biology courses, although more supervised laboratory sessions are required. Home-study kits for laboratories are supplied for chemistry and biology. They are picked up and turned in at the scheduled supervised labs. The kits contain basic laboratory items (e.g., test tubes, pipettes, etc.) and the student supplies other materials easily obtainable from around a typical household. The only courses that do not have supervised laboratories are geology and physical geography. The kits and activities that accompany the geology and physical geology courses can be completed entirely without supervision, due to the nature of the content area, i.e., rocks, soils, and land forms, as compared to working with living things and toxic chemicals. Computer simulations augment the astronomy course by using The Skyglobe™ software program. The software guides the students through specific astronomical simulations, therefore, eliminating the need for supervised laboratory experiences.

Many of the required supervised laboratories are offered at the AU campus or other post secondary institutions across Canada. These labs are typically made available during week-long sessions or weekend meetings throughout the year. Students report liking the extended lab meetings as they can make travel arrangements well in advance and complete the lab requirement in one session, or multiple sessions for biology and chemistry. AU is working toward infusing more electronic communications into these courses to bolster interaction between students and instructors, much along the line of the literature previously reviewed. Since AU is an open university, no comparisons can be made between on-site and distance learning science courses.

Kennepohl and Last's (1997) article was included in this review not because of its technological additions to the knowledge base on distance learning in science education, but rather due to AU's concern to offer science courses at a distance that require laboratory work. This one salient point is sorely missing from the many distance education science courses offered via the Internet as reflected in the literature.

The literature on the evaluation of distance education science courses largely uses commonly accepted and traditional methods of evaluation, e.g., test scores, that are typically used in on-site courses. Each reviewed study includes a component that tries to uncover student feelings and attitudes about the technological aspects of the instruction. Again, this type of evaluation methodology, for distance courses, is not without reason. One might believe that what is considered to be an accurate way of measuring student achievement in on-site courses can be easily exported and used with distance education courses as well.

An examination of distance education courses, and the accompanying research comparing on-site to on-line courses, or on-line courses by themselves, reveals the confusion about how best to evaluate the effects of such instructional experiences. The Institute for Higher Education Policy (TIHEP) (1999) cautions that, "Assessing the quality of the original research [on distance education courses] requires a determination that the studies adhered to commonly accepted principles of good research. The analysis is much more than an academic exercise. These principles are essential if the results of the studies are to be considered valid and generalizable" (p. 3). This one point, nevertheless, clearly defines the need for specialized evaluation procedures for on-line courses.

A Critical Review of Science Content Distance Education and Potential Alternatives

"When people think of the web," according to Fraser, "they usually think of it merely as a way to deliver long-established content to students at a distance. I treat it very differently" (Apple University Arts, 1999). Fraser is a professor of meteorology at Pennsylvania State University. His manuscript entitled *Colleges Should Tap the Pedagogical Potential of the World-Wide Web* (1999), which originally appeared in the *Chronicle of Higher Education* in August 1999, rebukes much of what has taken place in developing science content courses for distance education. He discusses the term "shovelware" in describing the phenomenon of how many contemporary college courses find their way onto the Web. Fraser (1999) states, "Shovelware can refer to any content shoveled from one communication medium to another with little regard for the appearance, ease of use, or capabilities of the second medium" (p. 2). He continues with:

> Today's new medium is the World-Wide Web. Quickly grasping its distribution
> possibilities, colleges and universities everywhere have rushed to move resources
> for courses on-line. Material previously handled on paper or with slides and
> transparencies—syllabi, assignments, notes, data, diagrams, references, exams—
> are now presented through the computer. (p. 2)

Fraser is one science content instructor who is heeding his own words in trying to unleash the potential of distance education. As an addition to posting *Introductory Meteorology* course materials on the web, he has developed diagrams, pictures, video clips, QuickTime VR movies, and animations. By adding these elements to his teaching, Fraser believes that key concepts in meteorology will be better understood by students. A typical class begins with Fraser, laptop underarm, getting connected to the Internet, and accessing various supplemental materials to enhance his lectures. "An important design criterion," according to Fraser, "is that these visualizations—plus everything else—must be available to my students in the lab, dormitory, and home in the same way that they are available to me in class" (Apple University Arts, 1999). What is the future of on-line learning? Fraser replies, "The near-term future is, alas, the continued delivery of shovelware ... How long will it take our pedagogical vision to move beyond the triviality of academic shovelware?" (Apple University Arts, 1999).

Adverse Effects of Distance Education

Even though from the published research into distance learning in all content areas one might think "all is well" when it comes to taking courses on-line, as with many stories, there are two sides. Hara and Kling (2000) recount one such story. They researched a Master's level on-line educational technology course offered in 1997 at a major university. The research design began as an ethnographic case study of student experiences (N = 8) while enrolled in the course, measured through interviews, on-line course observations, and a review of course materials. Five students were on-site; three were distance students. Only six students actually completed the course. Two of the distance students dropped the course due to technology problems. Many researchers have investigated and substantiated this fact (Burge, 1994; Gregor and Cuskelly 1994; Kang, 1988; Wiesenberg and Hutton 1995; Yakimovicz and Murphy 1995).

Hara and Kling (2000) found students experienced some degree of distress while enrolled in the on-line educational technology course. The authors define distress as "a general term to describe students' difficulties during the course, such as frustration, a feeling of isolation, anxiety, confusion, and panic" (p. 2). E-mail traffic was noted as a distress by the course instructor in terms of receiving some 35 messages about the class each day. Hara and Kling note that one student commented that just being able to talk about the course, face-to-face, would have been better. At one point, he devoted entire days to reading and responding to the course messages. This point is in stark contrast to Lennex's (2000) observation that, "Students have a distinct advantage in such courses because the instructor is much more available for assistance when the student needs it and not when the instructor happens to be in their office" (p. 5).

Chat sessions held during the course were also reported as problematic due to the students' inexperience with the technology. The conversations that were held sometimes scrolled too quickly across the screen and could not be read, reflected upon, and replied to in an appropriate manner. Many points of the discussions held were simply overlooked. Assignment requirements were also listed as a problem. Students reported that not having face-to-face interactions with the instructor, noting their body-language, led to misunderstandings. Working alone, during evenings or weekends, also distressed students. Hara and Kling (2000) conclude with:

> We suspect that the course's reliance on asynchronous communication further exacerbated the level of student distress over what would likely have been seen in a face-to-face class. Other game ecologies [computer-based interactions] have sometimes produced more satisfying instructional and personal experiences in distance education courses. We have much to learn about the conditions that create the good, the bad, and the ugly in Internet-enabled, text-based distance education. (p. 23)

In closing, the literature relating specifically to distance education and science content or science education courses is very limited. This point is somewhat understated as Hara and Kling (2000) report, "we found very few research reports" dealing with distance education research (p. 3). What does exist is a smattering of articles, many of which are not data based, that typically tell the reader of how others have taken an existing science content course and transformed it so that the course information, requirements, assignments, and tests are available on-line. Clearly, Kennepohl and Last's (1997) commentary regarding "no single solution" for exactly how to offer distance education science courses is well taken and supported by what little literature has been published (p. 37).

Evaluation of Distance Education: Potential Flaws in the Research

TIHEP (1999) reports, "In examining the more limited group of original research, several problems with the conclusions reached through the research are apparent" (p. 18). In particular, TIHEP discusses Russell's (1999) *The No Significant Difference Phenomenon*, a collection of diverse reports supporting the claim that distance education and face-to-face educational experiences are virtually equivalent in terms of student achievement.

As briefly noted earlier in this chapter, TIHEP (1999) cites the following shortcomings of distance education course evaluations and research:

1. Much of the research does not control for extraneous variables and therefore cannot show cause and effect.
2. Most of the studies do not use randomly selected subjects.
3. The validity and reliability of the instruments used to measure student outcomes and attitudes are questionable.
4. Many studies do not adequately control for the feelings and attitudes of the students and faculty—what educational research refers to as "reactive effects." (pp. 18–21)

The research articles reviewed in this chapter fall prey to TIHEP's (1999) concerns regarding the application of traditional research methodologies to very nontraditional classroom settings, namely on-line courses. The literature is most lacking in reports that demonstrate the efficacy of on-line courses. If any empirical evidence is offered at all, it usually manifests in anecdotal claims made by the researchers regarding overall student satisfaction in completing web-based courses. The reports of Ryan (2000) and Ingebritsen and Flickenger (1998), finding no significant differences between the final grades of on-line and on-site groups due to teaching techniques and styles appropriate for both course formats, miss the mark. The instruments used to gather data about students' similarities or differences enrolled in such courses were validated for use in educational experiences that were held in face-to-face [faculty-student] instructional settings.

A common misconception in educational research is that once an instrument has been found to be valid and reliable for the current group or groups under study, it then becomes valid and reliable, by default, for all groups administered the same instrument thereafter. While this action of "blind faith" is common practice, researchers should assess the validity and reliability of any instrument with each new group of participants tested or surveyed thereafter. This is a formidable task in light of the requirements that colleges and universities place upon course evaluations for each course taught during an academic year. The end result is that what might be a valid and reliable measure of course success and teaching effectiveness for one faculty member might not be the same for another. However, again, the common practice is to evaluate all faculty according to a standard measure.

As with any new educational innovation, the learning curve of such a novelty is only out distanced by an accurate system for rating its effectiveness. Distance education programs, especially in science content and science education courses, are two cases in point. Due to the very limited literature surrounding these cases, one would be hard pressed to even hint about the efficacy of distance education in science content or science methods courses, albeit for the positive anecdotal comments from students and instructors involved. This most important point must become the focal point of all distance education evaluation conducted in the future.

REFLECTIONS AND CONCLUSIONS

Although institutions of higher education are promoting and developing distance education opportunities at a staggering rate, the jury is far from reaching its verdict on what implications this exploding delivery system will have on science content and science education courses in the future.

For this writer, the case for offering distance education courses is a clear one. Institutions of higher education can enroll masses of additional students at little, if any, additional cost as compared to what the final cost would be to bring the same number of students on-site to attend classes. Tuition can be collected with virtually no impact upon the physical plant, impacting only the surrounding businesses that would stand to lose compensation for life's necessities. Graduate Assistants can be hired to help in the administration of an on-line course for a fraction of the cost of hiring new faculty to teach extra sections of the same. Students of all ages can exercise their "technological rights" by attending on-line classes at any time during the day and at any place where Internet access is available. They can work at their own pace, allowing for courses to be completed in a shorter period of time. This is in stark contrast to the traditional week by week class meetings held over the course of a semester or quarter. Class scheduling rules and procedures at many universities and colleges do not allow students to enroll in multiple on-site courses during the same time period each week. On-line courses, as well as time-tested correspondence courses, can be added effortlessly to a student's class schedule. Historically, colleges of continuing education have operated as separate entities within university systems, not being concerned with rigid class times and possible scheduling conflicts among all other courses offered on campus.

Clearly, not all students work at the same rate within a course. The potential, then, exists for various students to enroll in a heavier credit load, thus setting the stage for universities and colleges to offer even more on-line or on-site courses if the demand for additional course offerings arises. Important questions arise from this latent scenario. How valuable is the actual time expended in a course, aside from merely delivering the course content? Does the amount of time, while enrolled in a course, inhibit or enhance one's learning?

Frankly, it's an understatement to say, as did Venables (1998), that distance education, as a result of the exploding communications technology, is "here to stay" (p. 163). How far will we go down this path of distance education? Potentially, thousands of students could be enrolled in one course, given the proper amount of support, namely graduate teaching assistants to aid in the administration of a course. Two elusive questions remain—are students receiving the same type of educational experience as they would, had having face-to-face interactions with their peers and instructor? Might these on-line experiences be even more educationally valuable for our students?

Distance Education and Learning to Teach Science

The literature examining distance education science methods courses for pre-service teachers is nonexistent. To be clear, these courses are not solely based on delivering science content to students, but rather the courses designed to learn the pedagogy of teaching science. The current belief on how to teach science, in line with the spirit of the *National Science Education Standards*, is rooted in constructivist epistemology. Some authors have written about their attempts to base their content courses upon constructivist beliefs (Ingebritsen and Flickenger, 1998; Jonassen, Davidson, Collins, Campbell, and Haag, 1995). As Doolittle (1999) states, "It needs to be reemphasized that constructivism is a theory of knowledge acquisition, not a theory of pedagogy; thus, the nexus of constructivism and on-line education is tentative, at best" (p. 1).

One of the primary points in pre-service science teacher education is giving the students a chance to become learners themselves, often experiencing science in a way that was never part of their formal education, primary through college-level. The *National Science Education Standards* (1996) states that science is something that is often done to students, not done by them. Trying to break the habits of receiving science instruction in this manner is a formidable task. More than often, student teachers will slip into their prior knowledge mode and teach as they were taught, typically lecturing students, rather than trying out strategies influenced by constructivist epistemology.

How can distance education help pre service teachers learn new skills of instruction? The laboratory activities performed in science methods classes are often rated as the best part of the course. Students do not take on the role of children, but rather become active learners themselves.

Conversations, comments, observations, conjectures and questions flair from each working group. Wandering eyes catch a hint for the problem under study from another group. Again, more dialogue occurs within and between groups until the problem is solved. Data are shared. Clarification questions are asked. Content is learned as a result of the activity being performed first. For any college professor, the excitement never dims when students demonstrate that they have truly *learned* something.

Watching an instructor on television, or as streaming video on a computer monitor, is a current practice in distance education. If the sole purpose of the course is to teach content knowledge, then the medium used appears to have little influence on the learner (Clark, 1983). When trying to offer a holistic shared experience, between many people, distance education courses present new challenges to the instructor. Spector, Burkett, Barnes, and Johnson (2000) describe the experimentation of offering a web-based Science, Technology, and Society (STS) secondary science methods course in Florida. They state, "The class met seven times face-to-face out of the 15-week semester: five times on campus and two times at a model middle school that implements STS. Plans for the future . . . include testing the web site as a distance learning course with no on campus meetings" (p. 3). Questions arise as to how the course authors will offer, to their future students, the same experiences resulting from the face-to-face interactions, and, perhaps more importantly, the shared experiences of visiting a real school with real children, and watching model lessons being presented by a real, and not virtually based, middle school science teacher.

It is inevitable that some form of distance education experience will be substituted for real-time, real-life educational experiences. Many science content courses now offered via the web, offer credit without any form of practical laboratory work (Holmberg and Bakshi, 1982; Kennepohl and Last, 1997). The final questions for science content and science education instructors still remain—are such courses, without laboratory and practicum work, valuable? Do such on-line courses prepare students for advanced course work and professional lives? Do such courses prepare pre-service teachers for the realities of teaching science in the public schools?

With the explosion of this new medium, there assuredly will be a myriad of opportunities for instructors and researchers to design, implement, and assess the impact and efficacy of distance education programs upon the learning that takes place in science content and science education courses in the future. Therein lies the ultimate problem. How do we, as researchers and professional educators, assess the true impact of distance education upon our students? As with any traditional measurement, such as test scores, lingering concerns exist as to the validity of such evaluation tools, both from academic and professional perspectives. For example, institutions of higher learning have been using scholastic aptitude tests and grade point averages for a long time as a method of enrollment management, based upon the predictive value the measures have on the student's potential academic success during his/her first year of college. In the majority of instances, these predictions bear out. There are, however, students who score poorly on tests and prove to be less than stellar while enrolled in high school, that do, in fact, flourish academically.

A similar flaw exists in distance education programs. Even if students can successfully demonstrate their scholastic abilities via traditional testing methodologies, administered either on campus or on-line, does this truly demonstrate their potential success as a science teacher or professional scientist? At the very least, with on-site, face-to-face interactions between faculty and students, if it appears that the student needs additional support, the faculty member can intercede. Proponents of distance education maintain that this same level of service to students currently exists. Fraser (1999) might suggest that this form of formative and summative evaluation, used within distance education courses, is yet another example of "shovelware", whereby one type of evaluation of content mastery is merely shoveled from one communication medium to another.

This concern opens up an exciting field of possible research endeavors. Researchers have the opportunity to develop new evaluative frameworks and tools designed especially for measuring the efficacy of on-line courses, far above that of test scores and anecdotal data, often relating only to the ease, flexibility, and possible frustrations of completing courses on-line. Previously, students were bound to two forms of instruction—on-site or correspondence with little, if any, interaction with their professors.

With the continuing evolution of distance education programs, new areas of research may include comparisons of courses offered both on-site and on-line, investigating deeper levels of influence upon student achievement and performance. Potential research questions may emerge searching out truths and beliefs such as the influence of the professor, or lack thereof, upon the student within the context of distance education. Often, professors exude a tremendous effect upon the students by merely being an exemplar role model of a scientist or science teacher. Students learn by doing and watching effective practices take place. Will watching a streaming video of a professor teaching, or a scientist performing investigations in a laboratory, ever be as consequential as actually witnessing such behaviors firsthand? On the contrary, might certain professors be more suitable and effective within the distance education context? One need not look farther than the effect of notable scientists and informal science educators such as Carl Sagan, Don Herbert (Mr. Wizard), and more recently, Bill Nye, The Science Guy, upon learners of all ages.

The content of science and science education courses rer.1ains fairly stable, albeit the new discoveries added to the knowledge base with each passing day. Research involving the efficacy of content offered via the Internet and technology is without question. Will it be revealed that certain forms of content are more easily and effectively learned via distance education than on-site, face-to-face instruction?

The effects of the delivery system as reported by Clark (1983), regarding the instructional design versus presentation medium debate upon learning, and the concept of "shovelware", by Fraser (1999), stand at the center of evaluation of distance education. Technology continues to impact both instructional design and presentation media in massive ways. Nonetheless, can the Internet and technology not only enhance, but also transcend established and effective teaching practices students currently enjoy? This single, and most important question, remains elusive at the present time and should serve as a beacon to guide future research efforts in distance education relating to science content and science education courses.

REFERENCES

Apple University Arts. (1999, Spring). *Introductory meteorology courses transformed by the web.* [Available on-line: http://www.apple.com/education/hed/aua0101s/meteor]

Burge, E. J. (1994). Learning in computer conferenced contexts: The learners' perspective. *Journal of Distance Education* 9:19–43.

Charp, S. (2000). The role of the Internet. *T.H.E. Journal* 27:8–9.

Clark, R. E. (1983). Reconsidering research on learning from media. *Review of Educational Research* 53:445–459.

Collins, M. (1997). Developing and running a WWW biology course. *The American Biology Teacher* 59:594–596.

Curriculum Administrator's Schooltone Magazine. (2000, March). *Verbatim, John Gage: Technology made easy.* Part II of II, pp. 4–5.

Doolittle, P. E. (1999). *Constructivism on-line.* 1999 On-line Conference on Teaching On-line in Higher Education. [Available on-line: http://www.tandl.vt.edu/doolittle/tohe/on-line.html]

Edelson, P. J. (1998, February 17). *The organization of courses via the Internet, Academic aspects, interaction, evaluation, and accreditation.* Paper presented at the National Autonomous University of New Mexico, Mexico City.

Fraser, A. B. (1999). *Colleges should tap the pedagogical potential of the World-Wide Web.* [Available on-line: http://www.apple.com/education/hed/aua0101s/meteor]

Gregor, S. D., and Cuskelly, E. F. (1994). Computer mediated communication in distance education. *Journal of Computer Assisted Learning* 10:168–181.

Hara, N., and Kling, R. (2000). Students' distress with a web-based distance education course. Bloomington, IN: The Center for Social Informatics, Indiana University. [Available on-line at: http://www.slis.indiana.edu/CSI/wp00-01.html]

Ingebritsen, T. S., and Flickenger, K. (1998). *Development and assessment of web courses that use streaming audio and video technologies.* Iowa State University. (ERIC Document Reproduction Service No. ED 422 859)

Institute for Higher Education Policy. (1999, April). *What's the difference? A review of contemporary research on the effectiveness of distance learning in higher education.* University of Pennsylvania. [Available on-line: http://www.ihep.com/PUB.htm]

Imel, S. (1998). Distance learning. *Myths and Realities*, ERIC Clearinghouse on Adult, Career, and Vocational Education. [Available on-line: Jonassen, D., Davidson, M., Collins, M., Campbell, J., and Haag, B. B. (1995). Constructivism and computer-mediated communication in distance education. *The American Journal of Distance Education* 9:7–26.

Kang, I. (1988). The use of computer-mediated communication: Electronic collaboration and interactivity. In Bonk C. J. and King K. (Eds.), *Electronic collaborators: Learner-centered technologies for literacy, apprenticeship, and discourse*, Erlbaum, Mahwah, NJ, pp. 315–337.

Holmberg, R. G., and Bakshi, T. S. (1982). Laboratory work in distance education. *Distance Education* 3:198–206.

Lennex, L. (2000, January). *Distance education for newbies.* Paper presented at the annual meeting of the Association for the Education of Teachers in Science, Akron, OH. [Available on-line: http://www.ed.psu.edu/CI/Journals/2000AETS/00file1.htm]

National Center for Education Statistics. (1997). *Distance education in higher education Institutions.* Post secondary Education Quick Information System. U.S. Department of Education. [Available on-line: http://nces.ed.gov/pubs99/condition99/indicator-31.html]

National Science Education Standards. (1996). National Academy Press. Washington, DC.

Rekkedal, T. (1994). *Research in distance education—past, present, and future.* NKI Foundation, Norway. [Available on-line: http://www.nettskolen.com/alle/forskning/29/intforsk.htm]

Russell, T. L. (1999). *The no significant difference phenomenon.* Chapel Hill, NC: Office of Instructional Technologies, North Carolina state University.

Saba, F. S. (1999). *Distance education: An introduction.* [Available on-line: http://www.distance-educator.com/portals/research_deintro.html]

Schulman, A. H., and Sims, R. L. (1999). Learning in an on-line format versus an in-class format: An experimental study. *T.H.E. Journal* 26:54–56.

Spector, B. S., Burkett, R., Barnes, M., and Johnson, J. (2000, January). *Inter-institutional efforts to develop a web based STS course.* Paper presented at the annual meeting of the Association for the Education of Teachers in Science, Akron, OH. [Available on-line: http://www.ed.psu.edu/CI/Journals/2000AETS/00file1.htm]

The Pennsylvania State University. (1999). *Distance education catalog.* University Park, PA. [Available on-line: http://www.worldcampus.psu.edu]

The University of Phoenix. (2000). *The University of Phoenix Home page.* [Available on-line: http://www.phoenix.edu/faculty/index.html]

Truman, B. E. (1995, March). *Distance education in post secondary institutions and business.* Unpublished manuscript. [Available on-line: http://pegasus.cc.ucf.edu/~btruman/dist-lr.html]

Venables, J. A. (1998). Graduate education on the Internet. *Physics Education* 33:157–163.

Wiesenberg, F., and Hutton, S. (1995, November). *Teaching a graduate program using computer mediated conferencing software.* Paper presented at the Paper presented at the Annual Meeting of the American Association for Adult and Continuing Education, Kansas City, MO.

About the Authors

James W. Altschuld is a Professor in Educational Research, Evaluation and Measurement and the Evaluation Coordinator for the National Dissemination Center for Career and Technical Education at The Ohio State University. His research interests focus on needs assessment, the training of evaluators, and various aspects of the evaluation process. He has published extensively on evaluation topics including two co-authored books on needs assessment. He has received awards at the university, state, and international levels for his work in the evaluation field.

Bernice T. Anderson is a Program Director in the Division of Research, Evaluation and Communication at the National Science Foundation. She provides direction and oversight for the monitoring, evaluation and research studies of K-12 rural, statewide, and urban systemic reform programs. She also helps with the management of the Division's research projects focusing on diversity/equity issues. She has earned education degrees from the following institutions: Norfolk State University (B.S. in Early and Elementary Education, 1974), The Ohio State University (M.S. in Early and Middle Childhood Education, 1976), and Rutgers University (Ed.D. in Science and Humanities Education, 1984).

Gerard T. Calnin has been a research associate of the Centre for Program Evaluation at the University of Melbourne in Australia and completed his doctorate at the University. Recently he has assumed a senior curriculum and leadership position in the schools in the Melbourne area.

John R. Cannon is an Associate Professor of Science Education at the University of Nevada, Reno. He has been interested in electronic communication and e-publishing since 1987. Dr. Cannon is the editor and publisher of the *Electronic Journal of Science Education*, the first totally electronic journal devoted to publishing issues related to science education.

Dennis W. Cheek is Director of the Office of Research, High School Reform, and Adult Education at the Rhode Island Department of Elementary and Secondary Education and Adjunct Professor of Education at the University of Rhode Island. He earned a Ph.D. in Curriculum and Instruction/Science Education from the Pennsylvania State University and serves on the editorial or manuscript review boards of The Science Teacher, Journal of Technology Education, Middle School Journal, Odyssey, and Social Education. He has been active in numerous professional associations in education over many years and is a founding member of the Steering Group for the Campbell Collaboration, an international consortium devoted to systematic reviews of randomized controlled trials and quasi-experiments in education, social welfare, and criminal justice.

Douglas Huffman is an Assistant Professor of Science Education in the College of Education and Human Development at the University of Minnesota. He holds a Ph.D. in Science Education from the University of Minnesota with a minor in Evaluation, a M.Ed. from Harvard University in Education, and a B.S. from Stanford University in Civil Engineering. He is currently working on two NSF grants; a study of State Systemic Initiatives, and a national core evaluation of the Collaboratives for Excellence in Teacher Preparation. He also co-directs the Science and Mathematics Evaluation Fellowship Program designed to help Ph.D. students develop expertise in science and mathematics program evaluation.

David D. Kumar is a Professor of Science Education at Florida Atlantic University. His research involves evaluation and policy in science and technology education. He is the co-editor of the book Science, Technology, and Society: A Sourcebook on Research and Practice published by Kluwer Academic/Plenum Publishers (2000).

Faye C. Lambert has been a research associate of the Centre for Program Evaluation at the University of Melbourne in Australia and completed her doctorate at the University. Recently she has begun work in a senior curriculum and leadership position in the Melbourne area schools.

Wendy McColskey is the Director of the Assessment, Accountability, and Standards Program at SERVE (*www.serve.org*), which is a federally funded regional educational research and development laboratory serving the southeast located at the University of North Carolina at Greensboro. At SERVE for ten years, she has worked to develop publications and products that advance understanding and practice in the areas of classroom assessment, teacher evaluation, and district leadership of standards-based reform. Her Ph.D. is in educational research and evaluation from The Ohio State University. She co-wrote a popular SERVE publication entitled, *How to Assess Student Performance in Science: Going Beyond Multiple-Choice Tests*, with Rita O'Sullivan.

Rita O'Sullivan is an Associate Professor of Education at the University of North Carolina at Chapel Hill (UNC) and Director of Evaluation, Assessment, and Policy Connections (EvAP), a unit within the School of Education at UNC. She teaches graduate courses in Educational Program Evaluation, Case Study Methods, Research, Measurement, and Statistics. For the past 15 years, she has specialized in the evaluation of programs that promote academic success for diverse student groups and classroom assessment issues. In addition to her contributions to the field of evaluation and assessment via articles and presentations, she served as Secretary/Treasurer of the American Evaluation Association from 1992–1997.

John M. Owen is the Director of the Centre for Program Evaluation at the University of Melbourne in Australia for almost a decade. His interests are in effective knowledge use and innovative practices. He is the lead author of an evaluation text entitled "Program Evaluation: Forms and Approaches." In a previous life he had a major interest in science education.

Wolff-Michael Roth is the Lansdowne Professor of Applied Cognitive Science at the University of Victoria in Canada. He has published extensively in science education and has incorporated social, cultural, and psychological theory to address significant problems empirically and theoretically. His publications include numerous books and articles in the leading journals in science education, education, sociology of scientific knowledge, linguistics, and the learning sciences.

William H. Schmidt received his undergraduate degree in mathematics from Concordia College in River Forrest, IL and his Ph.D. from the University of Chicago in psychometrics and applied statistics. He carries the title of University Distinguished Professor at Michigan State University and is the National Research Coordinator and Executive Director of the US National Center which oversees participation of the United States in the IEA sponsored Third International Mathematics and Science Study (TIMSS). He was also a member of the Senior Executive staff and Head of the Office of Policy Studies and Program Assessment for the National Science Foundation in Washington, DC from 1986–1988. He is widely published in numerous journals including the Journal of the American Statistical Association, Journal of Educational Statistics, Multivariate Behavioral Research, Journal of Education Psychology, Journal of Educational Measurement, Educational and Psychological Science Measurement Journal, American Educational Research Journal and the Journal of Curriculum Studies and has delivered numerous papers at conferences including the American Educational Research Association, Psychometric Society, American Sociological Association, International Reading Association and National Council of Teachers of Mathematics. He has also co-authored seven books related to the Third International Mathematics and Science Study. He was awarded the Honorary Doctorate Degree at Concordia University in 1997 and most recently received the 1998 Willard Jacobson Lectureship from The New York Academy of Sciences.

Daniel L. Stufflebeam is a Professor of Education and Director of The Evaluation Center, Western Michigan University. He developed the CIPP Model, one of the first and still most widely used evaluation approaches; directed the development of more than 100 standardized achievement tests; and authored or coauthored 10 books and about 70 articles and book chapters. From 1975–1988 he chaired the Joint Committee that issued national standards for both program and personnel evaluations. For theory contributions, he received the 1984 Western Michigan University Distinguished Faculty Scholar Award and the 1985 American Evaluation Association Paul Lazersfeld prize. He is the co-editor and a frequent contributor to Kluwer Publishers' series of books on evaluation in education and human services.

Kenneth Tobin is Professor of Education in the Graduate School of Education at the University of Pennsylvania (Email: ktobin@gse.upenn.edu). Prior to commencing a career as a teacher educator, Tobin taught high school science and mathematics in Australia and was involved in curriculum design. After completing undergraduate and graduate degrees in physics, at Curtin University in Australia, he completed a doctorate in science education at the University of Georgia. His research interests are focused on the teaching and learning of science in urban schools, which involve mainly African-American students living in conditions of poverty. A parallel program of research focuses on co-teaching as a way of learning to teach in urban high schools. His recent publications include editing of the *International Handbook of Science Education* (with B. Fraser), *Re/Constructing Elementary Science* (with W-M. Roth and S. Ritchie), *At the*

Elbow of Another: Learning to Teach by Co-teaching (with W-M. Roth) and *Transforming Undergraduate Science Teaching: Social Constructivist Perspectives* (with P. Taylor and P. Gilmer Eds).

HsingChi A. Wang, Senior Researcher of US-National Research Center with Dr. William Schmidt on implications of TIMSS findings toward U.S. mathematics and science educational policy. Dr. Wang also is a Research Assistant Professor at the University of Southern California, where she has worked with Los Angeles Systemic Initiative on designing and evaluating science education reform projects. She received her bachelors degree in Physics in Taiwan, a masters degree in Science Education, and a Ph.D. in the Curriculum and Instruction of the University of Southern California School of Education. Her major research areas include the historical perspectives in science education reform, curriculum ideology versus science education standards, science instructional materials studies, science education program evaluation, and the relationship between mathematics and science learning. Furthermore, she also actively in research of Problem-Based Learning as it applied to various educational sectors. Since 1997, she has been actively engaging in the design of in-service workshops to bridge the gap between educational researchers and school educators using PBL approach.

Index